# Gunnar Uranium Mine

## Canada's Cold War Ghost Town

LAURIER L. SCHRAMM

**The Author:**
*Dr. Laurier L. Schramm*
Saskatchewan Research Council
125 – 15 Innovation Blvd.
Saskatoon, Canada, S7N 2X8

This book has been carefully produced. Nevertheless, the author and publisher do not warrant the information contained in this book to be free of errors. Readers are advised to bear in mind that the statements, data, illustrations, procedural details, or other items may inadvertently be inaccurate.

The views, opinions, assumptions, and estimates used or expressed in this book are those of the author and do not necessarily reflect the official policy or position of the Saskatchewan Research Council.

Copyright © 2016, 2017 Saskatchewan Research Council, Saskatoon, SK, Canada, S7N 2X8

All world-wide rights reserved, including those of translation into other languages. No part of this book may be reproduced in any form, by photo-printing, microfilm, or any other means, nor transmitted or translated into a machine language without written permission from the publisher. Registered names, trademarks, and the like, used in this book, even when not specifically marked or identified as such, are not to be considered unprotected by law.

Print ISBN: 978-0-9958081-2-6
ePub ISBN: 978-0-9958081-0-2

The watercolours shown on the cover and throughout this book are reproduced from Annual Reports of Gunnar Mines Ltd. for 1953, 1959, and Annual Reports of Gunnar Mining Ltd. for 1960, 1961, and 1962.

LAURIER L. SCHRAMM

## DEDICATION

All proceeds from the sales of this book will go to the Saskatchewan Research Council's Technology in Action Fund - a perpetual memorial fund established to help the people of Saskatchewan develop their province as a highly skilled, fair, desirable and compassionate society with a secure environment through research, development and the transfer of innovative scientific and technological solutions, applications and services.

# CONTENTS

| | | |
|---|---|---|
| | Preface | i |
| | Acknowledgments | iii |
| 1 | The Uranium Age | 1 |
| 2 | The Beginnings at Gunnar | 25 |
| 3 | Gunnar: The Town | 37 |
| 4 | The Gunnar Mine | 53 |
| 5 | Milling | 73 |
| 6 | Tailings and Waste Rock | 89 |
| 7 | Closure and Abandonment | 95 |
| 8 | Remediation: The Ending | 113 |
| 9 | A Final Accounting | 135 |
| 10 | Glossary | 139 |
| 11 | Appendices | 145 |
| 12 | Summary | 153 |
| 13 | About the Author | 155 |
| 14 | References | 157 |

# PREFACE

The Gunnar mine, mill, and town-site were built in a remote location in northern Saskatchewan, on the shore of Lake Athabasca, the 22nd largest lake in the world. Like most mining communities the town boomed, first with construction workers and miners, and later with families. When the Gunnar mill construction was completed in the fall of 1955 it doubled Canada's uranium production capacity. By 1956 the Gunnar mine was considered to be the largest uranium producer in the world.

The Gunnar mine produced over 5 million tonnes of uranium ore, nearly 4.4 million tonnes of mine tailings, and an estimated 2,710,700 m3 of waste rock. Following closure in 1964, the Gunnar site was abandoned with little remediation and no reclamation being done. One website refers to the Gunnar mine as "the second greatest environmental disaster area in Canada" [1].

The Gunnar town-site was built to serve the mine and mill and at one time had a population of about 850 people. By 1964 it was a ghost town. Ghost towns are not unusual in Western Canada, Saskatchewan has about 300 [2]. Of these Gunnar was probably the most recent[1], the largest at its peak, and possibly the most interesting.

Forty years would pass before the governments of Saskatchewan and Canada reached an agreement to fund the remediation (clean-up) of the Gunnar site, and contracted the management of the project to the Saskatchewan Research Council (SRC). At the time of writing this book the clean-up was well underway, with several years of clean-up activity remaining, and a further expected 10-15 years of monitoring activity before the site is expected to be released into a long-term management and monitoring program.

---

[1] Most of Saskatchewan's ghost towns were either founded by immigrant settlers in the 1800s and 1890s or in the early 1900s as railway lines were developed [2].

LAURIER L. SCHRAMM

## ACKNOWLEDGMENTS

Thanks to Ann Marie Schramm, William Schramm, and Dr. Joe Muldoon for reading and commenting on early drafts.

Special thanks also to Patty Ogilvie-Evans for her help finding old documents and references to Gunnar, and for reading and commenting on early drafts. Thanks also to Dr. Joe Muldoon for many helpful discussions about the old uranium mines and regulatory policies, and to Mark Simpson and Rostyk Hursky for supplying additional materials on Gunnar.

Even in the modern electronic and Internet age there remains a need for major research libraries with substantive collections of scientific and engineering books and periodicals. In the preparation of this book my work was greatly assisted by the collections of the libraries of McGill University, Massachusetts Institute of Technology (MIT), University of Saskatchewan, University of Regina, and the University of Toronto.

# 1 THE URANIUM AGE

## 1.1 Uranium Exploration – When it All Began. Time-Line: 1789 – 1937.

The mineral pitchblende was first identified and named in 1727 as the rock that glowed[2], at St. Joachimsthal in what is now Czechoslovakia [3]. This was before uranium itself was discovered by German scientist Martin (W.H.) Klaproth in 1789. At first pitchblende was simply a curious by-product of silver mining, and it seems to have been known in various parts of Europe by the late 1700s. It was later found to be a uranium oxide mineral (also known as uraninite) whose average chemical formula is $U_3O_8$. Some limited mining of pitchblende for use in colouring glass and porcelain [4] took place in Europe in the early 1800s (and probably in earlier centuries as well), but there was relatively little interest in the mineral until the discovery of X-rays by German scientist Wilhelm Röntgen in 1895, radioactivity by French scientist Henri Becquerel in 1896, and radium in 1898 by French scientists Marie and Pierre Curie (see Reference [5] and Figure 1.1). Radiation therapy for various diseases, particularly cancer, emerged shortly after the discovery of X-rays, and radium therapy became an even more popular method for radiation treatments beginning in the early 1900s. At about the same time radium came into industrial use as a luminous coating for glow-in-the-dark products, particularly instrument panels, watches, and clocks.

In the United States (U.S.), uranium ore was discovered in 1871 in gold mines near Central City, Colorado and later at the Colorado Plateau of Utah and Colorado. These areas were actively mined for their vanadium and/or

---

[2] The glow was probably due to radiation from uranium and radium causing zinc sulfide, which is also found in this particular pitchblende mineral, to phosphoresce [3].

radium contents in the late 1800s and early 1900s [6,7].

Industrial quantities of uranium, in the form of pitchblende, were first discovered at Shinkolobwe in the Belgian Congo[3] in 1915, as brightly coloured "*queer stone[s]*" [8]. There wasn't much industrial interest in uranium at that time but there was a market for radium[4], which is commonly associated with pitchblende. The Belgian company Union Minière du Haut Katanga mined the Shinkolobwe deposit for its radium, beginning in 1921 and continued for nearly two decades [3,8]. In the early years of radium therapy 100 mg of radium salt sold for about $12,000 (in 1918 U.S. dollars) [9]. When the market for radium severely weakened near the end of the 1930s, the Shinkolobwe mine was closed.

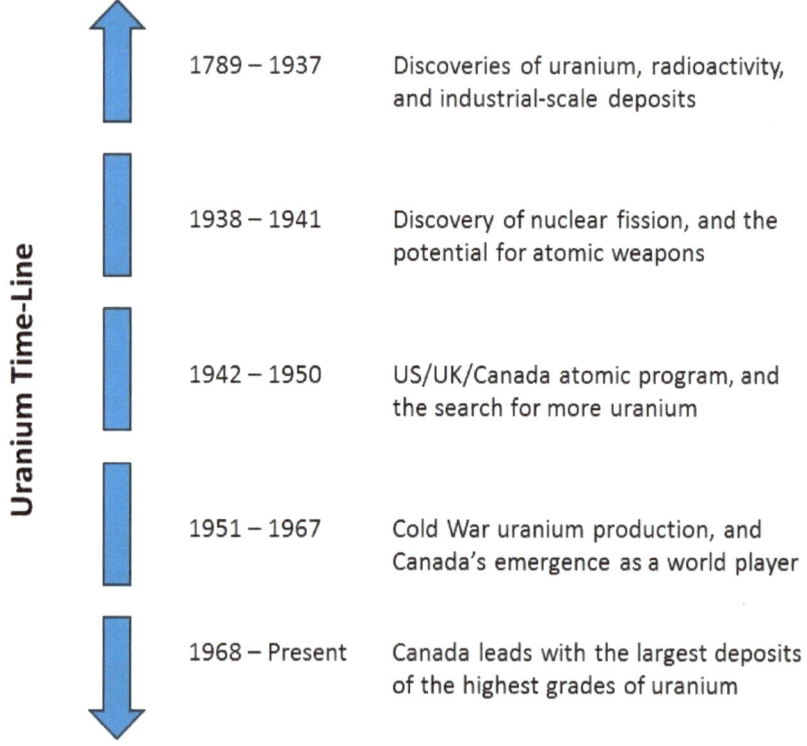

**Figure 1.1. A uranium development time-line.**

---

[3] The Belgian Congo achieved independence in 1960, and since then has been the Democratic Republic of the Congo.

[4] At this time radium was in demand for use in cancer treatments.

Canada's uranium story had begun somewhat earlier, when in 1900 two scientists from the Geological Survey of Canada, Mackintosh Bell and Charles Camsell, noticed brightly coloured *"lilac stain of cobalt"* on rocks along the shore of Great Bear Lake, Northwest Territories, [8,10]. Bell and Camsell didn't pursue their observations, but they did write a report on their findings. Nearly thirty years later prospector Gilbert LaBine studied their report as part of his research for upcoming prospecting expeditions. In the 1920's Gilbert and his older brother Charles LaBine had opened the Eldorado Mine in Manitoba, and they formed the Eldorado Gold Mines Ltd. in 1926 [11,12]. When the Eldorado Mine played-out Gilbert LaBine returned to prospecting and began searching near Great Bear Lake [13]. In 1930 LaBine, and his prospecting partner Charles St. Paul found Bell and Camsell's cobalt stains and discovered deposits of silver, cobalt, and pitchblende [8,10,11,14]. To develop this find they established a new Eldorado Mine near what would later become Port Radium[5] (see Figure 1.2). They began some mining in 1931/32 and brought the mine and a mill into continuous production in 1933 [10,12].

In order to refine the milled ore LaBine and St. Paul built a radium refinery at Port Hope, thousands of kilometres away on the shore of Lake Ontario in 1933 (see Figure 1.2). The Lake Ontario location was chosen in part because of its proximity to the suppliers of the large quantities of processing chemicals that were needed: about 7 tonne of chemicals were needed to process one tonne of uranium oxide concentrate [25]. By the late 1930s the new Eldorado Mine was a major international-scale producer of radium [8]. As was the case in Belgium, Canadian uranium was an unwanted by-product of the radium production. Nevertheless, LaBine decided to stockpile the uranium *"Thinking someday that it may be useful"* [11].

Uranium-bearing rock (pitchblende) was also discovered in several locations near the north shore of Lake Athabasca in Saskatchewan in 1934 through 1936 [10,15-17]. These were not followed-up with significant additional exploration and discoveries until 1944 when Eldorado Mining and Refining staked several claims in the Beaverlodge area (see Figure 1.5 below). Others followed suit in subsequent years. In 1949 uranium exploration in the area had dramatically increased and the abandoned town of Goldfields was revived [15]. Although thousands of radioactive "surface showings" had been discovered by this time, none would be developed until the beginning of the cold-war in 1951. In addition to developing the

---

[5] There were several name changes, first it was called Great Bear, then Cameron's Point in 1932 [3]. By 1933, with prospecting and mining on the increase in this area, the Canadian Government established a town named Cameron Bay, about 11 km south-east of present-day Port Radium. The government facilities in Cameron Bay were renamed Port Radium in 1936 and eventually the whole community came to be referred-to as Port Radium.

new Eldorado Mine and the Port Hope refinery, LaBine continued to explore in central Manitoba where, in 1934, he formed Gunnar Gold Mines, which remained in production for several years [18]. As will be discussed below, a second Gunnar mining company was to come later – in the 1950s.

Demand for radium exceeded supply throughout the 1930s, keeping prices high and making radium mining and milling quite profitable. By 1940, however, the adverse effects of radium on human health had become well known and its use in most medical treatments, consumer-health-products, and luminescent products had been discontinued. As a result, the radium market collapsed. This coupled with the existence of large inventories on hand caused the new Eldorado Mine to be closed in 1940 (as was the Shinkolobwe mine), and Port Radium also was virtually abandoned. However, both the mine and the refinery were destined to become revitalized only two years later as a result of the international atomic energy race.

Figure 1.2. Illustration of the location of Canada's first uranium mine - the new Eldorado mine and mill – near Port Radium, east of Great Bear Lake (1). Also shown is the location of Eldorado's uranium refinery at Port Hope on the north shore of Lake Ontario (2). The map itself approximates Canada as it was in the 1930s, and was drawn based on the "*Territorial Evolution, 1927*" map in the *Atlas of Canada*, 6th Ed. [19].

## 1.2 Nuclear Fission is Discovered – The Atomic Age Begins. Time-Line: 1938 – 1941.

In 1938, two German chemists, Otto Hahn and Fritz Strassmann discovered that when they bombarded uranium nuclei the nuclei split apart, yielding two approximately equal fragments of a lighter element, barium, but the mass of the fragments totaled to less than that of the original uranium [5,20]. Seeking to explain this phenomenon, two Austrian physicists, Otto Frisch and Lise Meitner developed a very important theory. They theorized that the nucleus of a uranium atom, when struck by neutrons, could be split into pieces – the smaller atoms observed by Hahn and Strassmann – plus neutrons, and a huge amount of energy. This process became known as fission. They published their theories in 1939 [21].

Subsequently Hahn, Meitner, and other physicists realized that such fission actually occurred for a specific isotope of uranium (U-235) that was normally present only in very low concentrations, typically less than one percent. However, if enough U-235 atoms were packed closely enough together, then the neutrons released by one atom could cause the breaking (or fission) of several other uranium atoms, which in turn would split apart releasing more neutrons, and so on, creating a very fast chain reaction and releasing an extraordinary amount of energy [5,22]. The potential to create such huge quantities of energy from very small amounts of material was both very exciting and very frightening.

While excited about the potential for a new source of almost unlimited energy, Frisch and Meitner immediately realized that such power could cause harm as well. Frisch wrote to the British government warning that a small piece of uranium could *"produce a temperature comparable to that of the interior of the sun. The blast from such an explosion would destroy life in a wide area ... probably cover[ing] the center of a big city"* [23]. Scientists in other countries, particularly in Germany, Russia, Canada [5,24][6], and the U.S. came to similar conclusions. Thus began a race to find and obtain uranium and to try to develop atomic weapons.

It had been estimated that the critical mass needed to enable the chain reaction in an atomic bomb would be about 10 kg (22 lb) of U-235, although the first-ever atomic bomb actually contained 64 kg (141 lb) [23]. Countries involved in the atomic weapons race naturally desired to stockpile as much uranium as possible as a strategic resource for their own

---

[6] In Canada, for example, Ernest Rutherford noted in 1904 that *"the total energy emitted from 1 gram of radium during its changes is about a million times greater than that involved in any known molecular change ... There is thus reason to believe that an enormous store of energy could be obtained from a small quantity of matter"* [24].

use. For a while it was thought that uranium was somewhat rare, with only three substantial deposits being known: Shinkolobwe in the Belgian Congo, Port Radium in Canada (see Figure 1.3), and St. Joachimsthal in a part of Czechoslovakia that had recently been annexed by Germany. The onset of the Second World War in 1939 heightened both strategic interests and fears, as it was also thought that the number of nuclear weapons could be limited by acquiring as much of the known uranium reserves as possible.

These factors set the stage for renewed uranium exploration and atomic power developments in Canada and elsewhere. The National Research Council built Canada's first laboratory-scale fission reactor in Ottawa in 1940 using uranium from Port Hope, via the Eldorado mine that had just closed [25].

Figure 1.3. Photograph of Port Radium in the 1930s. (Courtesy: Public Archives of Canada # C-23966).

## 1.3 Substantial Uranium Resources Needed – The Atomic Age Reaches Canada. Time-Line: 1942 – 1950.

Canadian exploration for uranium surged again beginning in 1942, due to military interest in building an atomic weapon. Eldorado had been selling their stockpiled, by-product uranium to the U.S. government in 1941 and 1942, but demand kept increasing [12]. In hopes of meeting the increased demand the Eldorado Mine was reopened in 1942 [25,26] and contracted to supply uranium to the U.S. Army, but it was clear that additional uranium reserves would also need to found [18]. The Eldorado refinery at Port Hope had remained in continuous production mostly producing radium, but by 1942 its focus had shifted to producing mostly uranium [3].

The Canadian government had imposed a ban on public prospecting for, and mining of, any kind of radioactive materials across Canada, but the government itself vigorously pursued these activities in secret [3,27][7]. In 1942 the United Kingdom (U.K.) and Canadian governments launched a joint atomic energy research program based in Canada [3]. By August 1943 the U.S., U.K., and Canada had merged their atomic weapons development programs under a cooperation agreement called *"The Articles of Agreement on Tube Alloys"* ("Tube Alloys" having been the code name for this project) [18,25]. This agreement was reached and signed in Canada at "The Quebec Conference," held at the Citadel in Quebec City. The "Tube Alloy" Project later became part of the Manhattan Project, led by the U.S. but still in cooperation with the U.K. and Canada. The world's first nuclear fission reactors were built in the U.S. and Canada. Part of Canada's role was to develop and build a heavy water reactor for the production of plutonium from uranium [25].

Included in the Manhattan Project was a component aimed at finding and acquiring as much uranium as possible, beginning in about 1942. At first, the uranium for the Manhattan Project came from the Shinkolobwe Mine in the Belgian Congo and from the Eldorado Mine (refined at Port Hope) in Canada. To this was later added uranium produced in the U.S. itself. Several American companies had been mining for vanadium in Colorado and Utah, but their ore actually contained both vanadium and uranium. The uranium in their ore had now become valuable, but to maintain secrecy the U.S. Army publicly maintained that they were only buying vanadium. At this point in time the largest known deposit of uranium was still at Shinkolobwe, but there was now a strategic reason to search worldwide for additional reserves: a desire on the part of the U.S., U.K., and Canada to control as much as possible of the world's uranium

---

[7] Attempts to keep Canada's uranium mining secret were not entirely successful. In July of 1943 Eldorado received an order for uranium from the USSR [25].

reserves. This was later enhanced by a growing interest in plutonium (made from uranium).

In 1942, the Canadian Government passed legislation reserving to the Crown ownership of all radioactive substances found in the Northwest Territories and Yukon. In order to maintain security the Canadian government also started purchasing shares in the Eldorado company [12,25]. In 1943 Eldorado Gold Mines Ltd. was renamed Eldorado Mining and Refining Ltd.[8] [3]. In 1944 the Canadian government expropriated all outstanding shares in the company and turned it into a Crown Corporation. Gilbert LaBine remained President of Eldorado through all of this, and had shifted the Eldorado Mine and Port Hope refinery focus to the production of uranium rather than radium. This involved converting uranium ore concentrate (yellowcake) into uranium black oxide (an orange-coloured solid comprising about 96% $U_3O_8$) [67]. For the next several years, this mine and refinery were the only significant new source of uranium in the western world. Also during this period, the Port Hope refinery received and processed ore from the Shinkolobwe mine [12]. Most of the uranium was sent to the U.S. and probably used to produce the first atomic bombs[9], including the first plutonium bomb ("Trinity") detonated at Jornada de Muerto, New Mexico in July, 1945 [11].

In 1946 the Canadian government established the Atomic Energy Control Board (AECB) to regulate the uranium industry and essentially all atomic energy activities [25]. In 1947 Canada began using Eldorado to stockpile uranium, in addition to supplying the U.S. [12]. Meanwhile, the Canadian government pursued additional secret uranium exploration and development activities across Canada using the resources of both Eldorado and the Geological Survey of Canada [14,27]. This wave of uranium exploration was aided by the availability of hand-held portable radiation detectors[10] [28,29]. The most significant finding of this relatively new wave was that the Beaverlodge region north of Lake Athabasca in northern Saskatchewan not only had pitchblende occurrences (as had been previously discovered in 1934) but had at least a thousand occurrences [3]. Of these, the first staking took place in 1944 [30] and the first large ore body was discovered in 1946 [3].

When the Second World War-era security restrictions were reduced in 1948 [8,14] private enterprises were again allowed to get involved in

---

[8] Eldorado Mining and Refining Ltd. later became Eldorado Nuclear Ltd.
[9] It has been estimated that about one-sixth of the uranium delivered to the US Manhattan Project came from Canada [7].
[10] Some of the first commercial hand-held radiation detectors include ionization-chamber detectors, such as the 1930s-era *"Curtiss Radium Detector"* and the 1940s-era *"Victoreen Model 247/247A,"* and Geiger-Müller counters, such as the 1930s-era *"Radium Hound"* [28,29].

exploration, mining, and milling, although all mined ores and concentrates were still required to be sold to Eldorado or other government-designated agency (and at a government-guaranteed price) [3,10,12,14,31]. The renewed exploration activities of 1948 and 1949 catalyzed the finding and development of new uranium deposits and about 45 small- to medium-size mines [32,33]. An example is the Madawaska/Faraday Mine near Bancroft, Ontario, which was discovered in 1949 but did not begin operating until 1957 [62] (see Table 1.1). In 1949 the only Western suppliers of uranium were still just Eldorado and Shinkolobwe but by 1950 both the U.S. and South Africa had also become significant producers of uranium. The price for uranium concentrate guaranteed by the Canadian government was increased in 1950 to encourage further exploration and the development of additional new uranium mines and mills [31].

The end of this era was also the beginning of the Cold War era. The pursuit of military and peaceful applications of nuclear energy, driven by both hope and fear, rekindled demand for uranium. By the end of the 1940s Russia was receiving all of the uranium ore produced by the St. Joachimsthal mine in Czechoslovakia, and by the Schlema mine in East Germany, and it had become known that Russia had successfully test-exploded an underground atomic bomb.

Canada had developed a series of research reactors during this era. The Zero-Energy Experimental Pile (ZEEP) Reactor was Canada's first nuclear reactor and the world's first non-U.S. reactor. It operated from 1945 to 1970 and was used to produce plutonium and uranium-233. Canada's second nuclear reactor, the National Research Experimental (NRX) Reactor, commenced operation in 1947 and remained in service until 1993. Meanwhile Eldorado had begun to sell cobalt-60, and by 1951 Eldorado, two Canadian university groups (in Saskatchewan and Ontario), and groups in the U.S. had developed cobalt-60 medical devices (called "cobalt bombs") to provide focused gamma rays[11] for radiation treatment of cancer [12,34]. By the early 1950s Canada had become the world's largest supplier of medical isotopes [35].

---

[11] Gamma rays are electromagnetic waves of very high energy (and very short wavelength). Gamma rays are one of the kinds of radiation that can be produced by radioactive atoms as they decay.

Table 1.1. Examples of Canadian Cold War-Era Uranium Mines. (Sources: [17,32,33,38,60].)

| Mine | Discovery | Mean Grade (% U) | Producing Years | Yield (tonnes $U_3O_8$) |
|---|---|---|---|---|
| Beaverlodge Mines and Mill, SK (Eldorado Ace-Fay-Verna) | 1946 | 0.24 | 1953-1982 | ~20,400 [62,66] |
| Cayzor Athabasca Mine, SK | ~1953 | 0.33 | 1954-1960 | 170 [16] |
| Cinch Lake, SK | 1948 | 0.20 | 1955-1960 | 259 [16] |
| Eldorado – Dubyna, SK | 1947 | 0.22 | 1978-1982 | 148 [16] |
| Eldorado – Eagle, SK | ~1946 | Erratic | 1950-1951 | 77 [16] |
| Eldorado – Fish Hook, SK | 1945 | 0.22 | 1957-1960 | 14 [16] |
| Eldorado – Hab, SK | 1958 | 0.43 | 1972-1976 | 692 [16] |
| Eldorado – Martin Lake, SK | 1946 | Erratic | 1948-1954 | 10 [16] |
| Eldorado Port Radium Mine and Mill, NWT | 1930 | | 1930–1940; 1942-1960 | |
| Gunnar Mine and Mill, SK | 1952 | ~0.2 | 1955–1963 | 6,252 [16] |
| Lacnor Mine, ON | 1953 | | 1957-1960 | 2.7 million |
| Lorado Uranium Mine and Mill, SK | ~1953 | | 1956-1960 | 81 [16] |
| Madawaska/Faraday Mine, ON | 1949 | | 1957-1964; 1975-1982 | 4,305 |
| National Explorations Keiller, Pat Mines, SK | 1951 | ~0.7 | 1954-1958 | 27 [16] |
| Nesbitt-Labine Uranium–Eagle, ABC, Mines, SK | 1950, 1952 | ~0.2 | 1952-1956 | 21 [16] |
| Nicholson Mine, SK | 1935 | ~0.4 | 1949-1959 | 37 [16] |
| Pronto Mine and Mill, ON | 1953 | | 1955-1960 | 2.1 million |
| Rayrock Mine, NWT | 1948 | | 1957-1959 | 207 [66] |
| Rix Athabasca – Leonard Mine, SK | 1951 | ~0.2 | 1955-1960 | 70 [16] |
| Rix Athabasca – Smitty Mine, SK | ~1949 | | 1952-1960 | 396 [16] |
| Uranium Ridges Mine, SK | 1950 | ~0.6 | 1958-1959 | 9 [16] |

## 1.4 The Cold War-Era Uranium Mines. Time-Line: 1951 – 1967.

With the beginning of the cold war the U.S. decided to continue its nuclear program, including expanding their nuclear arsenal and conducting research and development (R&D) aimed at developing a hydrogen bomb. These activities increased the demand for uranium from Canadian and U.S. mines. Russia had also decided to continue with its nuclear program, drawing uranium from Schlema in Germany and St. Joachimsthal in Czechoslovakia. The early 1950s also saw Britain independently continue its nuclear program, drawing uranium from Portugal, the Belgian Congo, and South Africa, and France launched a nuclear weapons development program as well. In Canada at this time, the Port Hope refinery became exclusively focused on producing uranium [3].

Beyond uranium-security and weapons programs, another factor contributing to demand for uranium was the emergence of nuclear power programs[12]. The United States had started a nuclear power program in the 1940s, and the first electric-power generating nuclear reactor, EBR-I[13], was built in Idaho and started-up in December, 1951 [23,36]. For its part, Canada had launched Atomic Energy of Canada Ltd. (AECL, a Crown Corporation) in 1952 and a nuclear power program in 1955, producing the nuclear power demonstration (NPD[14]) reactor, which was built in Ontario and started-up in June, 1962 [12,37]. The emergence of these nuclear power programs contributed to governments' desire to build uranium reserves, and regulatory and incentive changes by the Canadian and U.S. governments triggered uranium exploration rushes in both countries [3,6,12].

The regulated (Canadian) price for uranium was increased in 1950 and again in 1951 to stimulate exploration and to ensure that the Beaverlodge and other mines could proceed [31]. The Beaverlodge mine did proceed (beginning operations in 1953), as did a number of other, smaller mines [3,12]. In response to the "tent cities' that had begun to spring up around the individual mine sites, the Saskatchewan government established the community of Uranium City in 1951 with the aim of serving the entire region (see references [38-41]). Then, in 1952, the Saskatchewan government changed regulations making it more attractive for prospectors to explore and stake claims in the Lake Athabasca region (for which the Canadian Government would have exclusive purchase rights for all mined

---

[12] Industrial applications were also being developed, such as in nuclear density gauges and nuclear thickness gauges, but these did not significantly affect overall uranium demand.

[13] The EBR-I (Experimental Breeder Reactor I) produced 200kW of electricity and was operated from 1951 until decommissioning in 1964.

[14] NPD was the fore-runner of the Canada deuterium uranium (CANDU) power reactors.

uranium) and to support them [40]. These steps made it attractive for more companies, and even amateurs to prospect for uranium, triggering a massive uranium exploration and claim-staking rush [4,40,42-46] that helped Canada maintain its international position as a uranium producer[15]. A *Northern Miner* headline proclaimed *"Uranium - Canada Maintains Place in Frantic World Production Race"* [47] and a *Precambrian* article noted that *"Uranium deposits, it seemed, began to appear everywhere"* [48]. The Saskatchewan uranium rush even caught the attention of broadly circulated magazines like *Maclean's* and *Life* [49-52], and made headline news as far away as Australia [53-56] (see Figure 1.4). A television documentary film, *"The Birth of a Great Uranium Area,"* was made in 1953, illustrating the processes of uranium prospecting, drilling, and mining in the area [57]. By the fall of 1954 the government announced that 50 to 60 companies were actively engaged in uranium exploration, development, mining and/or processing in Northern Saskatchewan [58]. Uranium City itself grew to nearly 5,000 people (the size for which it had originally been designed [41]), see Table 2.1 below. Another television documentary film, *"The Road to Uranium,"* was made in 1957, illustrating life in Uranium City at its peak of 5,000 people, and of operations at the Eldorado mines and mill [59].

*The 1950s uranium rush attracted prospectors of all kinds, professional and amateur. In both Canada and the US, amateur prospectors could get basic prospecting instructions from inexpensive US Government pamphlets [45,46], maps from the Canadian and US Geological Survey Departments, and equipment from mail-order retailers like Sears Roebuck and Montgomery Ward [44]. In 1955 a complete prospecting kit could be purchased for US$3,529 that included everything from a Geiger-Müller counter to a Jeep [44].*

---

[15] At about the same time, related developments in the United States triggered a uranium rush there as well [6].

Figure 1.4. Illustration of Australian newspaper headlines reporting the 1952 Saskatchewan uranium exploration boom [53-55].

The increased level of exploration activity driven by the Cold War led to several significant Canadian uranium discoveries[16] beyond those in Saskatchewan, including deposits around the Bancroft, Ontario, area in the early 1950s (including the Madawaska/Faraday Mine mentioned earlier), and the first discovery in the Elliot Lake / Blind River, Ontario, region in 1953. This led to 12 mines including the Lacnor, Milliken, Nordic (Algom Nordic), Panel, Pronto, Quirke (Algom Quirke), Stanleigh, Stanrock, Can-Met, and Denison Mines [8,27,60] - see Table 1.1 and Figure 1.5.

Before World War II the cutoff grade for commercial production of uranium was about 0.5% (as $U_3O_8$), but the ability to apply more sophisticated processing technologies, and to work at larger scales of operation, brought the cutoff grade down to as low as 0.1 % by the mid-

---

[16] The search for additional uranium resources was not restricted to Canada, of course, and significant new deposits were found elsewhere during this period, such as at Mi Vida in Colorado in 1953.

1950s [8]. This change enabled many new mines to be developed in Canada and, again by the mid-1950s, several practical deposits had been identified [27] although some either did not get developed or did not last very long. For example:

- In 1949 uranium mineralization was discovered to be associated with silver and lead minerals in the Rexspar deposit in southern British Columbia. The deposit was further evaluated throughout the 1950s (and again in the 1960s and '70s) but was never brought into production [61].
- The Rayrock Mine near Great Slave Lake, Northwest Territories (see Figure 1.5) was discovered in 1948. Production at Rayrock began in 1957, but the mine only lasted for two years [3,62]. The mine and its small town-site were both abandoned in 1959 when its reserves were depleted [1,62].

There had been pitchblende prospecting and staking in Northern Saskatchewan since the 1930s and 1940s, but by 1952 there was enough uranium prospecting activity that the Saskatchewan government established the town of Uranium City, as noted above, to provide services to the mines in the Beaverlodge area north of Lake Athabasca (see Figures 1.5, 1.6). Several Saskatchewan discoveries followed that of Beaverlodge, including Radiore nearby, Rix Athabasca on Black Bay, and Nesbitt-Labine near Uranium City [12]. Gilbert LaBine and his son Joseph were part of the renewed prospecting in this region, and they found several deposits in the same general area in 1952 [8,11].

In July 1952 LaBine, Albert Zemel, and Walter Blair found and staked a uranium deposit at the southern tip of the Crackingstone Peninsula on Lake Athabasca [3,63] (see Figure 1.6). LaBine used his Gunnar Gold Mines Ltd. company to hold the title on the property, and dropped the word gold from the company's name to create Gunnar Mines Ltd. [12]. This was considered to be the richest uranium strike in Canada at the time [52], and Gunnar would become the first large private uranium mine of the era. The Gunnar Mine and mill site opened in September 1955, operating first as an open pit mine (1955–1961) and later as an underground mine (1957–1963) – see Chapter 2. When the Gunnar mill's construction was completed in the fall of 1955 it doubled Canada's uranium production capacity [64]. The Gunnar mine operated until 1963 and was the second highest producer of the 16 Beaverlodge area mines in the Atomic Age and Cold War Eras (Table 1.1; [65]).

Figure 1.5. Illustration of the locations of some of Canada's cold war era uranium mines: Eldorado Mine (1), Bancroft-area mines (2), and Elliot Lake - Blind River area mines (3), Rayrock Mine (4), Beaverlodge-area mines (5). The map itself approximates Canada as it was in the 1950s, and was drawn based on the *"Territorial Evolution, 1949"* map in the *Atlas of Canada*, 6th Ed. [19].

By this time Gilbert LaBine had become known as *"Mr. Uranium"* in Canada [52], having co-founded Eldorado, and its gold, radium, and uranium mines, the Port Hope refinery, Gunnar and its gold and uranium mines, and the Nesbitt-Labine uranium mine. He had stewarded Canada's uranium production through the World War II years, and now through the Cold War years, and was probably the most visible champion of Canada's uranium mining industry during these eras.

Through Eldorado Nuclear the LaBines began mining at their Beaverlodge site[17] in 1953, where they established a dedicated mill and also

---

[17] The Beaverlodge Mine has sometimes been referred-to as the Eldorado Mine, although as noted above there were two previous Eldorado Mines, one in Ontario and an earlier one in Manitoba.

a small community named Eldorado, all located about 7 km east of Uranium City [27,66]. The Saskatchewan Beaverlodge site, which included the Ace, Fay, and Verna mines[18], operated until 1982 and was the highest producer of the 16 Beaverlodge area mines of the Atomic Age and the Cold War Eras (Table 1.1; [65]).

The Lorado Mine opened in 1957 and operated until 1960 [62] (see Figure 1.6). Lorado was the only other mine to have its own dedicated mill (along with Beaverlodge and Gunnar), and the Lorado mill operated from 1957 to 1961. In addition to processing the mined ore from their own mines, the Lorado and Beaverlodge mills also processed ore from smaller mines in the region, including the Cayzor, Rix Leonard, and Cinch Lake Mines, and also ore from "surface miners" who picked over surface showings and waste rock piles that were uneconomic for the mining companies to handle [62,63].

In total, more than 1,000 pitchblende occurrences were discovered in Saskatchewan's Beaverlodge district, but only 16 mines of significant size were actually brought into production between 1953 and 1982 (see Table 1.1) [16]. By 1960 LaBine estimated that the Beaverlodge district mines had produced about $300 million worth of uranium (in 1956 dollars) [30].

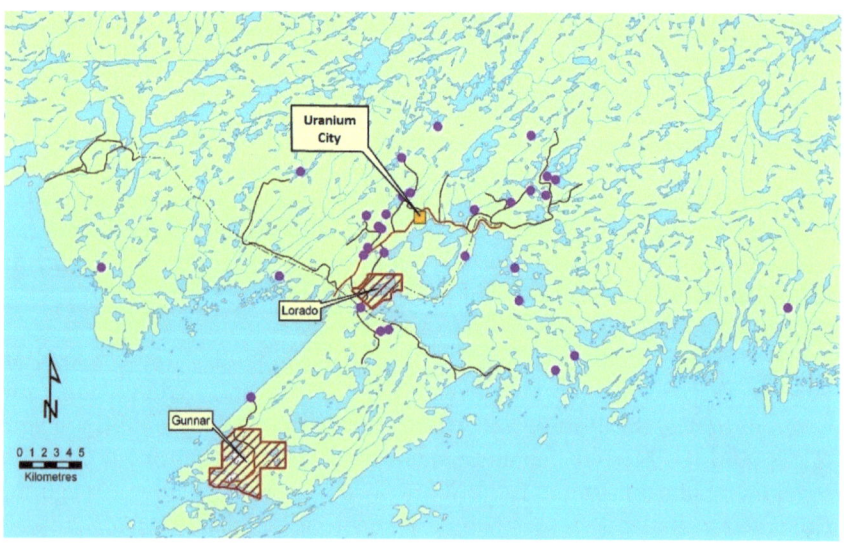

Figure 1.6. Illustration of the locations of two of Saskatchewan's cold war era uranium mines: the Gunnar and Lorado Mines.

---

[18] An illustration of the early days of establishing Eldorado's Ace and Fay mines is given in a TMC documentary [57].

By 1957 there were 18 operating uranium mines in Canada (see Table 1.1), all of which sold their uranium ore (raw or milled) to Eldorado, which in turn sold the uranium to the U.S. [12]. This number peaked at 21 Canadian producing mines in 1958, by which time 11 of the mines plus three mills were operating in the Uranium City area [3,14,62].

Canadian uranium production levels grew throughout the 1950s. In 1956 the "free-world" production of uranium was close to 13,000 tonnes [8]. Most of the production came from the U.S., Belgian Congo, and Canada (see Table 1.2). By 1958 half of the world's production of about 27,200 tonnes was estimated to have come from "*Arctic Canada*" meaning the Port Radium Eldorado Mine [8,27]. By 1959 uranium was Canada's number one mineral export (ahead of aluminum, iron, and nickel) with 23 mines and 19 mills in operation [27]. Of the 19 mills, 11 were in the Elliot Lake, Ontario area, three near Bancroft, Ontario, three in northern Saskatchewan, and two in the Northwest Territories [27].

**Table 1.2. World uranium reserves in 1957.**
(Conversions to S.I. units are approximate. Based on data in Reference [8].)

| Country | Production in 1956 (tonnes $U_3O_8$) | Uranium Reserves (millions of tonnes $U_3O_8$) | Average Ore Grade (mass%) |
|---|---|---|---|
| South Africa | 3,992 | 998 | 0.03 |
| Canada | 2,994 | 204 | 0.10 |
| United States | 5,443 | 54 | 0.24 |
| Total | 12,429 | 1,256 | |

At about this time radiation safety was becoming better understood. Although by 1940 the adverse effects of radium on human health had become well known and its use in most medical treatments and consumer products had been discontinued, the health effects of low doses of radiation were not yet well known. The concepts of safe working levels and safe cumulative (annual) exposure levels emerged in the 1950s, although uranium mining in the 1950s was still considered "safe" (as far as radiation hazards were concerned) [25]. In 1959 Gordon Churchill, the federal Minister responsible, confidently stated in the House of Commons that

"... *there are no special hazards attached to the mining of uranium that differ from other mining activities*" and "... *there is no radiation hazard in the processing operations*" [25].

Nevertheless, in 1960 the AECB created regulations dealing with radiation safety [25]. The 1960 AECB regulations defined for the first time the concept of an "*atomic energy worker*" and the maximum amounts of ionizing radiation to which such a worker could be allowed to become exposed[19].

Most (about 90%) of the Canadian uranium produced in the 1950s and early 1960s was sold to the U.S. Atomic Energy Commission, with most of the remainder being sold to the U.K. Atomic Energy Authority [3,10,14]. The Rayrock mine closed in 1959. Then the Eldorado Mine closed in 1960, when the uranium ore ran out, although the Eldorado mill was able to continue operating until 1967. Despite depleting ore bodies, by this time uranium production was outstripping demand. By 1962-63 the U.S. had more than enough uranium for its needs and the U.S. Atomic Energy Commission began reducing its purchases [67]. As a result, the levels of both exploration and mining and milling decreased, and the number of active mines shrunk to only five, the Madawaska/Faraday, Milliken, and Nordic mines in Ontario, and the Beaverlodge and Gunnar mines in Saskatchewan [14,27]. A 1966 Saskatchewan Department of Mineral Resources report concluded that "*uranium was a glut on the market, and without markets incentive is lacking to search for and develop new mines*" [14]. For a while the Canadian government supported the uranium industry with a stockpiling program, but this only lasted until 1974 [27].

Although there was an oversupply of uranium in the markets, the late 1960s were also characterized by much eager anticipation of nuclear power and clean energy. As already noted, first power-generating nuclear reactor, EBR-I, had begun operating in Idaho, but not many other power reactors had yet been built. Nevertheless, with forecasts that the world's known reserves of uranium[20] could be quickly used-up if many nuclear power plants were constructed [14], prospecting for new deposits increased again, and once again Saskatchewan was "*now one of Canada's busiest [uranium] prospecting areas*" [14].

Uranium production in the Bancroft and Beaverlodge areas ended in 1982, and in the Elliot Lake area in 1996 [27]. Thus ended the second era of Canadian uranium production. The first two eras of Canadian uranium

---

[19] These regulations also defined the maximum amounts of ionizing radiation to a member of the general public could be allowed to become exposed, at 1/10th of the amount for an atomic energy worker.

[20] In 1964 the known world reserves of uranium were 430,000 tonnes, 40% of which was located in Canada (Saskatchewan Department of Mineral Resources, 1966).

production helped put Canada on the world stage from an industrial point of view, and these eras are rich in stories. One of these is the story of the Gunnar Mine, which is described beginning in Chapter 2.

The Beaverlodge area mines and smaller developments of the Atomic Age and the Cold War Eras left behind an unfortunate legacy in that they were simply abandoned without much or any cleanup, and frequently without significantly closing-off the various mine shafts, adits, and rises. A 2006 Saskatchewan Environment and Resource Management report identified 45 such abandoned mines in the immediate Uranium City area alone [33]. These aspects are described further, in the context of the Gunnar operation, in Chapter 8.

A third era would emerge in 1968 driven by the need for uranium to fuel nuclear power reactors. This third era would see larger mines, much higher ore grades, and more varied mining and milling technologies on the strength of which Canada would once again be a major international producer (see Section 1.5).

## 1.5 Modern-Era Uranium Mines. Time-Line: 1968 – Present.

There was a significant lull in Canada's uranium production between the end of the cold-war era and the early 1990s. This was partly due to market conditions, as described above, but also partly due to changes in the political and regulatory environment. Although the Atomic Age and the Cold War Eras (1938 through 1967) exhibited uranium mine developments across Ontario, the Northwest Territories, and Saskatchewan (see Table 1.1), the modern-era uranium developments in Canada all took place in Saskatchewan.

In 1982 the Saskatchewan government had changed its policy on uranium and decided to phase-out existing uranium mines and to prohibit new ones. This policy lasted until about 1992, at which point the province reversed course and once again became an active supporter and co-developer of uranium mining in Saskatchewan [68].

Whereas the early Saskatchewan developments were in the Beaverlodge area immediately north of Lake Athabasca, the next boom in uranium exploration - in the 1970s - resulted in huge uranium discoveries in the Athabasca Basin area immediately south of Lake Athabasca. With these discoveries, from the early 1990s through 2008 Saskatchewan became the uranium capital of the world with the highest productions levels and the largest deposits of the highest grade of uranium on the planet. By 2009 the number one producer had become Kazakhstan, with Canada (Saskatchewan) in second place.

All of Canada's modern-day uranium production comes from mines in northern Saskatchewan (see Figure 1.7) and, in a way, Eldorado Nuclear is still involved. That's because the federal Crown Corporation Eldorado Resources Ltd. was merged with the provincial Crown Corporation Saskatchewan Mining Development Corp. in 1988 to create the Canadian Mining and Energy Corporation, commonly known as Cameco (see Table 1.3). Cameco in turn has been a major player in the majority of Canada's modern-era uranium mines.

Saskatchewan also has one nuclear reactor in operation. The Saskatchewan Research Council has operated a SLOWPOKE 2 research reactor in Saskatoon since 1981 [69].

**Table 1.3. Evolution of Eldorado Companies.**

| Name | Dates | Notes |
|---|---|---|
| Eldorado Gold Mines Ltd. | 1927 – 1944 | Gold mining (Manitoba); uranium mining (Northwest Territories); uranium refining (Ontario) |
| Eldorado Mining and Refining Ltd. | 1944 – 1968 | Uranium prospecting and development; uranium mining (Saskatchewan) |
| Eldorado Nuclear Ltd. | 1968 – 1988 | Uranium production for power plants |
| Cameco Corp. | 1988 - Present | Merged with Saskatchewan Mining Development Corp. to become Cameco (Canadian Mining and Energy Corp.) |

Another big difference is that these modern-era developments involved incredibly rich ore zones (see Table 1.4) and enabled a huge increase in Canadian uranium production. Although the average ore grades for the richest uranium mine in the world, the McArthur River mine is an average grade of 16.4% $U_3O_8$, during mining operations individual pockets of uranium ore have been found with uranium contents as high as, for example: 70% at the Cluff Lake mine [70]. By 2008 more than ten times as much uranium had been produced from the Athabasca region as was produced during the entire operating history of the Beaverlodge region (322 thousand tonnes vs. 30 thousand tonnes as $U_3O_8$) [65].

The first of the modern-era mines was the Rabbit Lake Mine, which was discovered in 1968. Opened in 1975, it was the longest operating uranium production facility in North America, with the second largest uranium mill in the world until its suspension in 2016 [65,72].

**Table 1.4. Canada's Modern-Era Uranium Mines.**
(Sources: [33,70-72,74-78].)

| Mine | Discovery | Average Grade (mass% uranium) | Producing Years | Yield (tonnes $U_3O_8$) |
|---|---|---|---|---|
| Cigar Lake Mine, SK | 1981 | 18.3 | 2014 to Present | 130,000 (projected) |
| Cluff Lake Mine and Mill, SK | 1971 | 0.6 | 1980 to 2002 | ~28,125 |
| Key Lake Mine and Mill, SK | 1975 | 0.52 | 1983 to 2002; Milling Continues | ~94,350 |
| McArthur River Mine, SK | 1988 | 16.4 | 1999 to Present | 175,000 (projected) |
| McClean Lake Mine and Mill, SK | 1979 | 2.2 | 1995 to Present | ~22,680 by 2015 |
| Rabbit Lake Mine and Mill, SK | 1968 | 0.70 | 1975 to 2016 | ~91,626 |

The Cluff Lake Mine was discovered in 1971 and operated from 1980 to 2002. The facilities included both open pit- and underground mines, plus a mill, tailings area, residential camp for employees, and other infrastructure [73]. Over its working life this mine produced about 28,350 tonnes of ore having an average grade of about 0.6% uranium [73].

The Key Lake Mine was discovered in 1975, operated from 1983 to 2002 producing about 4 million tonnes of ore having an average grade of 0.52% uranium. Although the mine is closed the mill, which is the world's largest high-grade uranium mill, continues to operate, being fed ore from the McArthur River mine [74].

The McClean Lake Mine was discovered in 1979, and has operated several open pit mines on the property in phases dating from 1995 to the present time. The ore being mined at this site has an average grade of about 2.2% $U_3O_8$. The McClean Lake Mill is also being fed with ore from the Cigar Lake Mine.

The Cigar Lake Mine was discovered in 1981, and has the world's second largest high-grade uranium deposit, with grades that are 100 times the world average [75,76]. This mine began production in 2014 and is expected to reach "full" production in 2018. The ore from Cigar Lake is transported to the McClean Lake site for milling.

The McArthur River Mine was discovered in 1981, and is the world's largest high-grade uranium mine [77,78]. It has operated from 1999 to the present day. The average ore grades in this deposit are 100 times the world average at about 15.8 % $U_3O_8$. The ore from McArthur River is transported to the Key Lake site for milling.

Canada's overall production was about 9,000 tonnes of uranium per year in the early 1990s and has been about 12,900 tonnes of uranium per year from the late 1990s to the present [79].

According to the World Nuclear Association the world's total recoverable resources of uranium in 2011 were 5,327,200 tonnes (uranium-basis) with about 31 percent in Australia, 12 percent in Kazakhstan, and 9 percent each for Canada and Russia [80]. Of this, a total of 58,394 tonnes (uranium-basis) were produced in 2012, about 36.5 percent by Kazakhstan, 15.4 percent by Canada, 12 percent by Australia, 8 percent each by Niger and Namibia, and 5 percent each by Uzbekistan and Russia [81].

***The Modern-Era Need for Decommissioning and Remediation.***
While the end of the Cold War changed the nature and extend of nuclear developments worldwide, this and the Chernobyl accident of 1986 changed attitudes towards nuclear safety and environmental protection. As a result many nuclear facilities established since the 1950s became redundant, many others reached the end of their design lives, leaving behind large areas of contaminated facilities and land [82]. According to the International Atomic Energy Agency (IAEA), "*Many countries were therefore left with facilities requiring to be decommissioned and/or sites requiring to be remediated*" [82][21]. Canada is no exception.

---

[21] See the glossary for definitions of the terms "*decommissioning*" and "*environmental remediation.*"

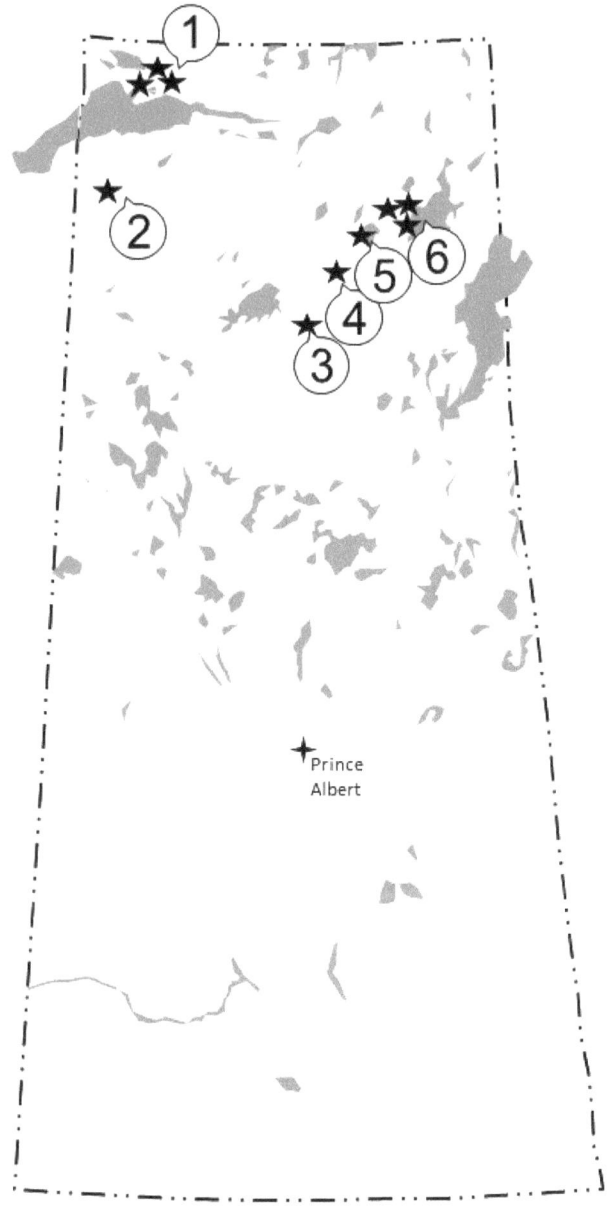

Figure 1.7. Illustration of the locations of Saskatchewan's cold war- and modern-era uranium Mines: Gunnar, Lorado, and Beaverlodge mines (1); Cluff Lake mine (2); Key Lake mine (3); McArthur River mine (4); Cigar Lake mine (5); and Midwest (proposed), McClean Lake, and Rabbit Lake mines (6). The city of Prince Albert is shown for reference. Drawn based on the "*Saskatchewan*" map in the *Atlas of Canada*, 6th Ed. [19].

# 2 THE BEGINNINGS AT GUNNAR

As discussed in Chapter 1, all of the conditions necessary for the development of several uranium mines in the Beaverlodge area of northern Saskatchewan came together in the 1950s, one of which was the Gunnar Mine. The cold war had increased demand for uranium and a combination of technology advances and the 1950-1951 price increases had made low-grade (~0.1 %) ore deposits economic.

The Gunnar deposit was discovered by Albert Zeemel in 1952. According to his account of events [83], Zeemel had been sent by LaBine to the "Athabaska" area to prospect for uranium, although *"I did not know too much about ... this uranium prospecting business. In fact, I didn't know anything about it."* Zeemel spent some time at the Nesbitt-LaBine uranium mine learning about uranium prospecting. He said that he learned to *"Always travel slowly and keep your geiger[22] on"* until *"I felt somewhat justified that I had a good idea of what I should be looking for and what I should like to find"* [83]. Travelling with other prospectors John Nesbitt, Pat Riley, and Walter Blair they would set up camp in a likely-looking area, search around - independently and working for different corporate interests – then compare notes on their findings, and move on the next likely location. Moving progressively southwest from Beaverlodge Lake they each found, staked, and sampled interesting areas until one evening Zeemel was following an interesting fracture zone when *"my geiger really began to buzz."* Finding additional hot zones nearby Zeemel sampled and staked the area, and spent several days completing a radiometric survey of the surrounding area[23]. The area Zeemel staked would become known as the Gunnar uranium deposit, at the

---

[22] Geiger counter.
[23] An illustration of the process of uranium prospecting in this area and time period is given in a TMC documentary [57].

southern tip of the Crackingstone Peninsula on the northern shore of Lake Athabasca (Figures 1.6 and 2.1) [3,63].

Returning to Edmonton Zeemel sent a radiogram to his boss Gilbert LaBine saying *"Come quick. I've shot an elephant,"* the term "shot an elephant" being code for making a uranium strike [84]. LaBine immediately flew out from Ontario and when Zeemel showed him the site and samples concluded that *"It looks big ...,"* and after spending more time at the site, *"Gosh, Albert, it looks bigger yet, it's just terrific"* [83]. Zeemel reported that a drill was next shipped to the site and eleven holes were drilled, the results from which *"indicated that we really had something that had the earmarks of a large mine"* [83]. By this time winter was rapidly approaching and their last task of the season was to arrange for a larger drilling program to commence in January of 1953 [83]. Within two years the mine would be in production.

LaBine used the dormant Gunnar Gold Mines Ltd. company to hold the title on the property [85], and dropped the word gold from the company's name to create Gunnar Mines Ltd. in 1954[24] [12,86]. The Atomic Energy Control Board granted a mining permit in 1954[25] authorizing Gunnar Mines Ltd.:

*"to carry on development, mining, milling and concentrating operations on the property hereunder mentioned and to ship ore and/or concentrates from there to Eldorado Mining and Refining Limited (hereinafter called "Eldorado") in accordance with such arrangements as may from time to time be in effect between you and Eldorado."*

A unique feature at Gunnar was the ability to begin with an open-pit mine, which had much lower startup costs than the underground mines. These factors led to Gunnar being the next major Canadian uranium mine development following the Beaverlodge Mine. Gunnar was also considered to be the richest uranium strike in Canada at the time [52], and Gunnar would become the first large private uranium mine of the era. When news of the uranium strike began to spread, the Gunnar share price *"jumped from forty cents to twelve dollars a share"* [84]. While some people rushed to buy Gunnar shares, others were just as quick to play down the find. Competitors reportedly referred to the discovery as having been found in a "sneer zone" meaning an area in which the geological formation was such that *"only a fool would bother to prospect [there]"* [84]. Its detractors notwithstanding, Gunnar was destined to become the second highest producer of the 16 Beaverlodge area mines of the Atomic Age and Cold War Eras (Table 1.1; [65]).

---

[24] The name of the company was later changed yet again, to Gunnar Mining Ltd., in 1960.

[25] Mining permit MP2/54, Sept. 16, 1954. Another one was issued, MP 2/57, on Feb. 1, 1957.

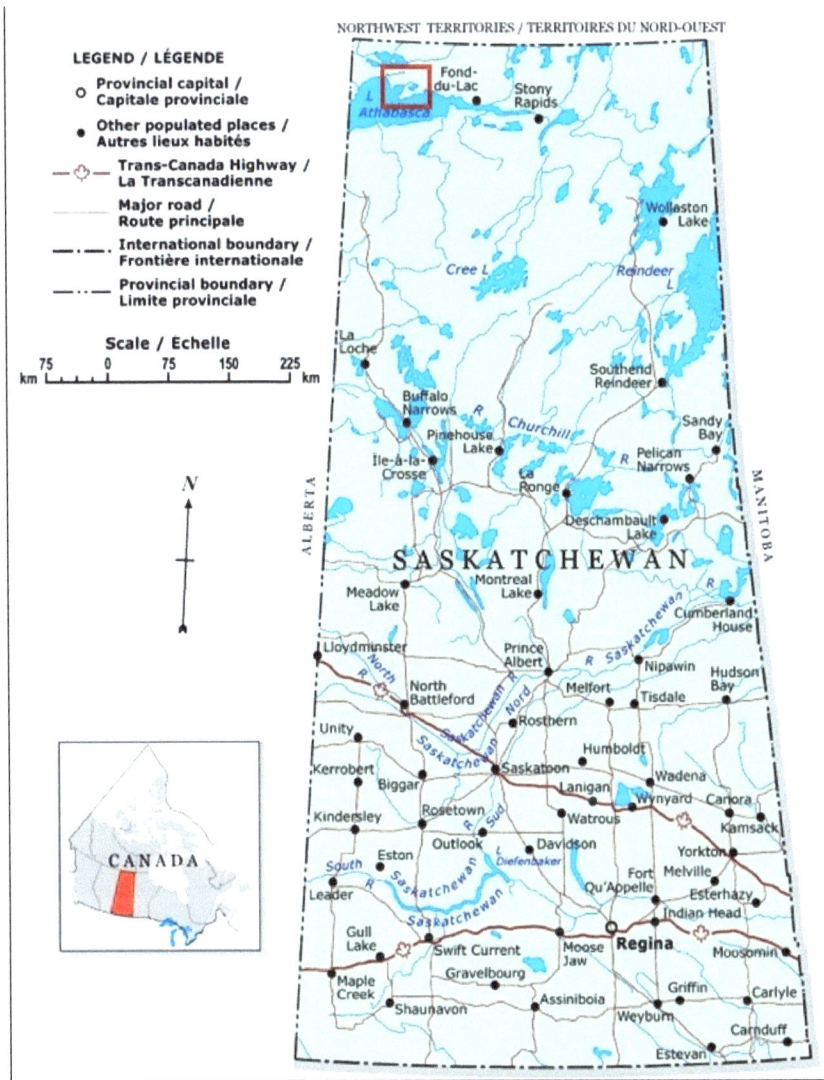

Figure 2.1. Gunnar mine and mill site location (red square in extreme upper left corner of the map. (Map from Natural Resources Canada, 2001.)

Although the Gunnar ore body is usually described as being "near" Uranium City, it is actually about 40 km away and was considered to be too far away for routine services, so it was decided to build a mill and a dedicated community in addition to the mine.

As already noted, diamond-drilling was commenced late in 1952 and expanded in early 1953, very soon after Zeemel's discovery of the deposit, [10,85]. From 1952 through 1954, stock markets and the media followed the progress of drilling at the Gunnar site with much interest. Newspaper articles in Toronto and New York extolled the latest drilling results and speculated on the development of the Gunnar site [87-96] (see Figures 2.2, 2.3), and drilling successes at Gunnar fueled exploration and development interests at nearby sites as well [97-100]. By April of 1953 Gunnar was beginning to mine a 27 tonne (30 ton) sample[26] to be shipped to the federal Department of Mines in Ottawa and used to establish a recovery process [101]. By June of the same year Gunnar was estimating that it would cost about $7/tonne to process their ore[27], which they estimated was worth about $27/tonne. On the strength of these estimates a 680 tonne/day (750 ton/day) mill had been ordered [102].With the mill on order, construction of a dock and warehouse next to the lakeshore were begun in July of 1953, and plans were already underway to add offices and housing on-site [103]. By early 1954 the company had already contracted with Eldorado Mining and Refining Ltd. to purchase uranium concentrates once the mine and mill were built and commissioned [86]. According to Hunter [31] the agreement with Eldorado by which the government would commit to purchase a large amount of high-grade uranium concentrate, at a fixed minimum price, and for five years, was the first contract of its kind in the industry.

In 1953 and '54 the sense of a booming industry in Northern Saskatchewan, and announcements of the forthcoming Gunnar mine and mill, again caught the attention of world media in broadly circulated magazines like *Maclean's* and *Life* [49,84,104,105], and in headline news reported as far away as Australia [106,107] (see Figure 2.4).

Being in a remote location, and without road access, most construction and other materials had to be brought in by boat or barge, or else by air (beginning in 1954 with aircraft landing on the lake, when frozen, and later using the Gunnar airstrip). This continued to be the case throughout construction, the operating lifetime of the mine, mill, and town-site, and through the final cleanup process. In the 1950s the cost of freight by air was more than double that by water so a year's worth of supplies would be ordered before the beginning of the boat/barge season[28] for delivery in advance of winter freeze-up [3]. The company originally used Eldorado's barge services [108], but eventually developed a dedicated Gunnar tug and barge system for shipping goods about 440 km (265 miles) along Lake Athabasca and the Athabasca River from the railhead at Waterways, Alberta

---

[26] Ultimately a 54 tonne sample was shipped to Ottawa for this purpose [110].

[27] Their cost estimates (1954) for mining the ore were about $2/ton (open pit) and $4/ton (underground).

[28] Ice breakup would usually occur in late May, with some areas not clearing until mid-June, and the lakes would usually be frozen-over by the end of October yielding an effective barge season of about 15 weeks each year [15,30].

[10,109]. In 1953 3,000 tonnes of freight was transported by barge from Waterways to Gunnar, increasing to about 20,000 tonnes in 1954, and over 30,000 tonnes in 1955 [108]. As a "rule of thumb" it was found that both capital and operating costs for the Athabasca region mines and mills, including Gunnar, tended to be more than double those of comparable southern operations [3].

Figure 2.2. Illustration of 1950s news headlines reporting on drilling results at Gunnar [87,93,94].

By August of 1953 enough bush had been cleared to make room for the mill and initial camp buildings, and the original cookhouse and 50-person bunkhouse were already having to be expanded in order to accommodate an increasing construction workforce at the mine [110]. Two additional 80-person bunkhouses and a 200-person cookhouse were built at this time. By

October construction had begun on machine, carpenter, and mechanical shops plus additional offices and warehouses [111].

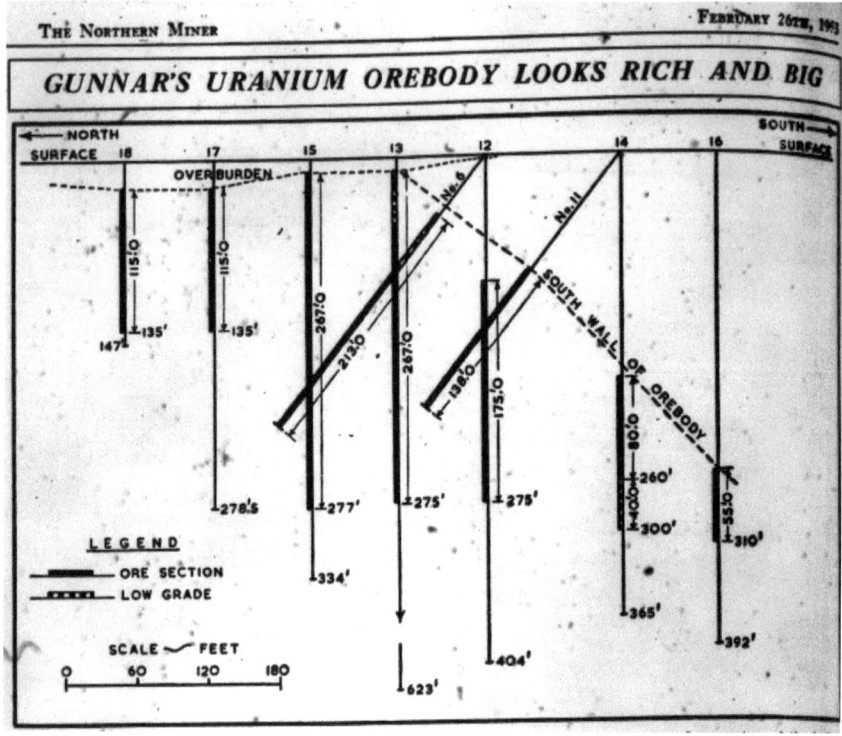

Figure 2.3. Illustration of 1953 diamond drilling results from the Gunnar deposit [90].

By the spring of 1954 the company name was changed to Gunnar Mines Ltd. [86] and a new wave of development activity was unleashed. Diamond drilling continued but most of the activities were focused on the mill and camp infrastructure. By March 1954 the heavy equipment, the acid plant, and a diesel power plant were all on order and contracts had been negotiated for the construction of roads and an airstrip [112] (see Figure 2.5). Contracts for sulphur (for making sulphuric acid for the leaching process) were established with Shell Oil Co. and Royalite Oil Co.[29] [86,113]. Although plant startup was still more than a year away the drilling program showed that the orebody was larger than had originally been thought, and plans were already under development to increase the size of the mill from 680 tonne/day (750 ton/day) to 1,134 tonne/day (1,250 ton/day) [86].

---

[29] The sulphur came from Jumping Pound, Alberta and was a by-product of Alberta natural gas production.

With constantly increasing estimates of the size and quality of the ore body financing for all of this was reported to be "no obstacle" [112] (see Figure 2.6).

Figure 2.4. Illustration of 1950s Australian news headlines[30] reporting the Saskatchewan uranium mine development boom [106,107].

In April 1954 the federal Department of Mines announced that it had completed its metallurgical testing on the Gunnar ore and that an acid-leaching process had been tested that would achieve "over 90% recovery" [114]. Based on the bench- and pilot-scale testing, the company selected a sulphuric acid-sodium chlorate digestion process and an ion exchange-magnesia precipitation separation/concentration process [128] (see Chapter 5). With the ore-body sufficiently well defined, and the camp expanded to 350-person capacity, a contract was established for the stripping of the ore body that would mark the beginning of the open pit mine [114]. Mining Permits were issued by the Atomic Energy Control Board (AECB)

---

[30] The Rum Jungle referenced in the right-hand article was a uranium deposit in the Northern Territory, Australia. Discovered in 1949, a mine and mill were constructed in 1952 that operated from 1953 to 1971.

authorizing Gunnar Mines Ltd.:

> *"to carry on development, mining, milling and concentrating operations on the property hereunder mentioned and to ship ore and/or concentrates from there to Eldorado Mining and Refining Limited (hereinafter called "Eldorado") in accordance with such arrangements as may from time to time be in effect between you and Eldorado."* (MP2/54 issued September 16, 1954, and MP 2/57 issued February 1, 1957 by AECB)

**Figure 2.5. 1954 photo of a dragline working on airstrip construction [86].**

Over the 1953-54 winter large planes, a DC-3[31] and a Bristol Freighter, were used to airlift heavy construction equipment to the Gunnar site, including bulldozers, scrapers, dump trucks, conveyors, compressors, a power-shovel, a crane, a large crusher, boiler, and a concrete mixer [115]. (Another DC3 aircraft was later to become a regular part of the Gunnar operation - see Chapter 5.) By June 1954 the on-site workforce had grown to 250 and accommodations for up to 500 were being planned, most of the shop buildings had been finished, and airport and mill construction were well underway [116].

---

[31] Another DC3 aircraft was later to become a regular part of the Gunnar operation (see Chapter 5).

**Figure 2.6. A 1955 Gunnar Mines Ltd. stock purchase warrant.**

The rapid construction progress continued into 1955. By January the mill and acid plant buildings had been completed and most of the equipment had been installed, the million-dollar (1955 dollars) power plant was almost complete, and construction of the town continued with the first dozen single-detached houses being completed and another dozen under construction (see Chapter 3) [117,118]. By May 1955 the company reported that all of the plant buildings had been completed, the power plant had commenced operation, and it had already removed nearly 600,000 tonnes of overburden rock [119,120] (see Chapter 4). By June the 1,830 m (6,000 ft) airstrip had been completed [121]. Near the end of August 1955 the mill construction had been completed, the acid and mill plants had been started-up, the first ore was fed to the mill, and the first yellowcake product was produced in September [10,122-124,128]. Figure 2.7 provides a sense of the location and footprint of the mine, mill, residences, and airstrip, stretching from St. Mary's Channel to Langley Bay.

By 1955, the Gunnar town-site population had grown to 510 (see Table 2.1) supported by some stores, the hospital, 5 large bunkhouses, 15 houses and more accommodations still under construction [118,125-127]. The town-site would continue to grow to its 800-person capacity [128] and beyond, as noted in the next chapter.

By 1957 the mill capacity had been expanded to 1,500 tonne/day, and the company reported making subsequent improvements that enabled it to achieve over 1,900 tonne/day in later years [3]. Figure 2.8 illustrates the breadth of the mine, mill, and town-site developments that had been completed by 1957. Appendix 2 shows a pictorial drawing of the site from 1961.

**Figure 2.7. Gunnar mine and mill site location (the larger black squares at the bottom of the map). The airstrip location is shown in the upper right.**

With a fully mature operation established Gunnar Mines Ltd. began telling its story publicly, with a booklet published in 1957 ([64], Figure 2.9) and an article in the Canadian Mining Journal in 1963 [128].

In 1960, Gunnar Mines Lt. and Nesbitt LaBine Uranium Mines Ltd. amalgamated under the new name of Gunnar Mining Ltd. [130].

Figure 2.8. A 1957 photo of the Gunnar mine, mill, and town-site. From Reference [129].

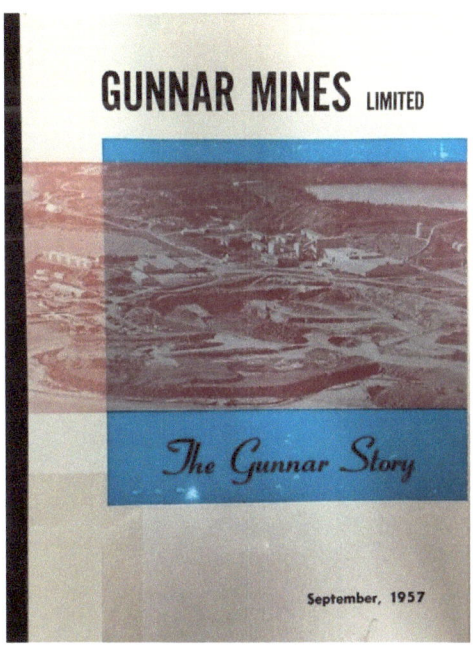

Figure 2.9. "The Gunnar Story" booklet, 1957.

**Table 2.1. Population Growth at Uranium City and the Gunnar Town-Site.** (By 1964 the permanent population at the Gunnar town site was reduced to a single caretaker.)

| Year | Uranium City and District | Gunnar Town-site | References |
|---|---|---|---|
| 1952 | 1,461 | | [39] |
| 1953 | 1,899 | 50-200 | [39,110] |
| 1954 | 2,500 | 250-450 | [39,58,116] |
| 1955 | 3,500 | 510 | [39,118,125] |
| 1956 | ~3,700 | | [39,42] |
| 1957 | 4,400 | | [39] |
| 1958 | | 800 | [129] |
| 1959 | 4,600 | 850 | [39,42,131] |
| 1960 | | 767 | [130] |
| 1961 | | 779 | [132] |
| 1962 | 3,000 | | [133] |
| 1964 | | 1 | |
| 1966 | ~2,100 | | [39,42] |
| 1969 | ~2,000 | | [17] |
| 1971 | 2,149 | | [39] |
| 1974 | 2,000 | | [39] |
| 1976 | 2,191 | | [39] |
| 1979 | 2,558 | | [39] |
| 1981 | ~2,500 | | [40] |
| 1982 | 800 | | [39] |
| 1986 | 200 | | [40] |
| | | | |
| ~2015 | ~60 | 0 | |

# 3 GUNNAR: THE TOWN

## 3.1 Establishment of the Gunnar Town-Site.

The site was (and still is) somewhat remote, existing in a sub-arctic region that is semiarid with short cool summers and cold winters. Being located on the shore of a large lake in Saskatchewan's north, the site was subject to almost constant winds, cool temperatures in the summer, (averaging 17 °C in July, with a maximum recorded high of 35 °C in 1984), and very cold temperatures in the winter (averaging -27 °C in January, with a maximum recorded low of -49 °C in 1974) [64,134]. The closest settlement to the site was Uranium City. Some descriptions of Uranium City and life in the town in this time period are given in references [1,38-42], and illustrated in the 1953 TMC [56] and 1957 ITN [59] documentary films.

As noted in Chapter 2 the company intentionally did not build a connecting road to Uranium City. In addition to Uranium City being an inconvenient distance away, the company also believed that easy access to the city would only create what mine manager Foss Irwin called "new problems" for them to handle [131]. As a result, the company developed and managed a complete town-site capable of supporting a population of up to 800 people [3]. Access to the Gunnar mine, mill, and town-site continued to be restricted to travelling by air or over water (or ice). The town-site was also managed as a "dry town," even to the point of checking taxis in the winter when returning from Uranium City over the ice road to ensure that any passengers were sober [135]. The company was able to attract a workforce to such a remote location by paying close to the highest wages in the mining and mineral industry in Canada [128].

By 1955, the Gunnar town-site population had grown to 510 (see Table 2.1) supported by some stores, the hospital, 5 large bunkhouses, 15 houses completed, and another 34 houses and a 16 suite apartment building under construction [118,125,126]. In 1956 the apartment building was completed, another bunkhouse was added, the school was completed, and the shopping centre construction was well underway [127,136]. By the fall of 1956 many families had already moved-in and classes had begun at the Gunnar school [136]. Canadian "Group of Seven" artist A.Y. Jackson visited the Gunnar site in 1957 and made several graphite drawings of the site, two of which are held in the collection of the National Gallery of Canada and can be viewed on their website [137,138].

The community shopping centre was completed in 1957, and by 1958 a seventh large bunkhouse had been constructed. The town-site ultimately included homes (Figure 3.1), dormitories, a hospital, community centre, school, and recreational facilities [3,128]. It would continue to grow to its 800-person design capacity [128] and beyond.

Up to 497 single men could be accommodated in seven two-story men's bunkhouse dormitories (Figures 3.2 and 3.3). Single women were accommodated in rooms in a women's residence above the hospital, and also above the dining hall. Up to 118 families could be accommodated in 86 single-dwelling houses or in two apartment buildings of sixteen suites each, all of which were supplied fully furnished [128,131]. The "West Townsite" was somewhat like a small modern-day suburb, comprising about 60 of the houses in a residential setting (Figure 3.3, upper), and located about 0.6 km (0.4 miles) west of the school, apartment blocks, and recreation centre. Appendix 1 shows an early surface plan of the site.)

**Figure 3.1. Sketch from the original plans for a senior staff house. Gunnar Gold Mines Ltd., 1954.**

The two-level, 3700 m² (40,000 ft²) community centre was centrally heated, and was also North America's first covered shopping mall. It included both shopping and community facilities. The shopping facilities included a Hudson's Bay department store, Imperial Bank of Canada branch, post office, bakery, butcher shop, barber shop, beauty parlor, dining hall, coffee shop, and more [128,131,135,139,141]. The stores sold food and supplies at cost, a practice that helped the Gunnar residents but irritated store owners in nearby Uranium City [142]. Black soil was even barged in to permit gardening [139]. The company also subsidized the housing, covering about two-thirds of the actual cost [131], and it provided all utilities, radio communication facilities, and transportation services (for both people and supplies) [3].

Figure 3.2. 1954 (upper and middle) and 1955 (lower) photos of Gunnar town-site residences. From References [86,118].

Figure 3.3. 1958 image of the growing number of single-family residences in the "West Townsite" (above-right) [139]. 1960 photo of the mine manager's house, which was located very close to the mine itself (lower-left) [140].

Radio communications were not just for recreational use however. Company President Gilbert LaBine related a story that illustrates the importance of radio at the time. According to his account [30], in February of 1956:

*"An accident occurred ... [a] workman received serious head injuries at 10:30 pm on Saturday necessitating immediate attention and evacuation to a city hospital for neurosurgical attention. Gunnar radio was able to communicate with Uranium City ... [but not] with any outside contact and it was only through an amateur radio man on the ham bands by way of Calgary and then Grand Prairie were we able to contact Edmonton by the early hours of Sunday morning. The RCAF search and rescue were dispatched and evacuated the patient to Edmonton, but not through any normal communication."*

The community facilities included a Y.M.C.A., gymnasium, bowling alleys, billiard room, library, lounge, camera club, hobby craft room, radio broadcasting studio, and more [128,131,135,139,141]. The gymnasium had a large adjoining stage and was designed so that it could easily be converted to an auditorium or theatre with a seating capacity of 700.

**Figure 3.4. The community centre in 1958 [139].**

**Figure 3.5. Shopping (above), bowling (middle), and gardening (lower) at the Gunnar site in 1958 [139].**

A dedicated school was constructed in 1956, containing three classrooms and a science laboratory [64]. Two more classrooms were added in 1958 [128] and at about this time there were four teachers. The student and teacher populations varied year by year, but in 1962 the Gunnar School's first yearbook listed a Principal and six other teachers covering kindergarten and grades 1 through 10, and a student population of 119 students [143].

The community included a seven-bed hospital with a resident doctor and four nursing staff [128,143]. First aid and out-patient services were provided on-site, while most surgeries were referred to a medical clinic in Uranium City, and/or Edmonton.

Fire protection was provided by a volunteer fire department, dedicated fire truck (Figure 3.7), and a high-pressure fire water delivery system spanning the entire site [128].

Additional recreational facilities included a three-sheet curling rink, outdoor ice rink, sports field, and a marina and beach on Lake Athabasca [128] (Figure 3.8). The company had barged-in sand from the massive sand dunes on the south shore of Lake Athabasca, and used it to build the beach in front of the town-site [135]. There were also two groupings of at least 5 and 19 cabins near the site [144]. For those wanting to get away from the town site, there was a picnic/camping area on a small lake about 5 km (3 miles away) [128].

According to former residents the Gunnar town-site was a good place to live, "everyone knew everyone else," and "people were united by isolation and a common purpose" [135].

**Figure 3.6. The Gunnar school (to the right) in 1958 [139]. The curling rink is shown to the left.**

Figure 3.7. The Gunnar fire truck in 1958 [139].

Figure 3.8. The Gunnar marina on Lake Athabasca, in 1958 [139]. A nearby beach was built by barging sand across the lake from the great Athabasca sand dunes.

## 3.2 General Infrastructure.

Since virtually the entire site was solid granite, all buildings and houses were connected by above-ground wooden pipe boxes called utilidors. These contained pipes for domestic water, fire water, and steam supply, and also sewage and effluent return pipes, some of which were asbestos-wrapped. With rare exception no pipes were run underground. The utilidors were placed at ground level and/or supported on wood trestles. The utilidors were insulated with wood shavings which, together with heat from the steam pipes, kept the pipes from freezing.

Lake water was treated with sodium hexametaphosphate, to reduce hardness, and piped to a 909 kl (200,000 imperial gal) water storage tank, which was located on a hill immediately behind the mill to provide elevation above the entire site (see Figure 3.10, below). From the storage tank, up to 18 million litres of water per day (4 million imperial gal per day) were supplied to the mill and town [63]. Domestic water was chlorinated.

Power for the entire site was originally generated in a single power house building by four 1,200 hp diesel generators (three more were added later) direct-coupled to 2,300 volt generators [128,145] - see Figure 3.9. The diesel engines were fueled with "Bunker B" type heavy diesel oil, which had to be heated and centrifuged prior to use. Typically, one generator would be kept on standby with all others in continuous operation. The same powerhouse was equipped with air compressors and vacuum pumps which provided high and low pressure air and vacuum for mine, mill, and shop use. Also in this building were two oil-fired boilers which were used to generate saturated steam for plant heating.

**Figure 3.9. The diesel power plant. Gunnar Mines Ltd., 1957 [64].**

About 10,000 tonnes of bulk petroleum products were shipped to Gunnar during the open-water season each year. These included heavy diesel fuel, light diesel oil, and gasoline. These were stored in large tanks located at the edge of the lake (Figure 3.10).

Steam heat was provided from three sources, the two boilers in the powerhouse, excess steam from the acid plants, and waste heat from the power and compressor diesels. Steam from these sources was combined and fed to a common low pressure steam system for the entire Gunnar site (mine, mill, and town).

Further details, including descriptions of the company's administrative functions, can be found in references [64,128].

Figure 3.10. The fuel storage tanks are shown in the lower left (along the lake shore) and the water tower is shown in the upper right. Gunnar Mines Ltd., 1958 [139].

## 3.3 Transportation Infrastructure.

The preferred means of transportation was by water since Lake Athabasca and the Athabasca River connected the Gunnar site with the community of Waterways, Alberta, which in turn was served by the Northern Alberta Railway. This provided about six months of ice-free waterway per year, but only about four months per year were practical for waterborne transportation in light of high winds and low water levels each fall season [64,128].

Waterborne freight was moved by barges towed by shallow-draft tug boats using either the Northern Transportation Company (a federal Crown Corporation) or the company's own Gunnar Water Transportation Division, which was created in 1956 and comprised a tug and six 400 tonne-capacity barges [146,147] (see Figures 3.11, 3.12, and 5.10). Although the operating seasons for the barge were only about 15 weeks each year [30], in Gunnar's peak operating years some 35,000 tonnes of freight per year were transported to the site by these barges [139]. Even the barge operations experienced challenges, however. For example, according to Quiring [148], *"a boat towing eight barges loaded with material for Gunnar and Eldorado froze into the ice in 1956."*

Figure 3.11. The first Gunnar-owned tug. Gunnar Mines Ltd., 1957 [64].

Figure 3.12. A Gunnar Water Transportation Division barge approaching the Gunnar dock in 1958 [139].

Air transportation was also critical to the construction and operation of the Gunnar mine, mill, and town. The aircraft first used to fly yellowcake out to Eldorado were Canadian Pacific Airlines DC-3s, [149]. In 1953, Gunnar-Nesbitt Aviation Company (a subsidiary of Gunnar Gold Mines Ltd.) was incorporated to provide dedicated air transportation services. The air service operated between Edmonton, Beaverlodge, and Gunnar. At first, a Mark V Anson aircraft was purchased, but this was replaced in 1954 with a Douglas DC-3C conversion (originally a U.S. Army Air Force C-47A-DK), which was purchased by Gunnar Mines Ltd. and Nesbitt Labine Uranium Mines Ltd. (Figure 3.13). Once the mine and mill were in production, yellowcake from Gunnar was transported by DC-3 to Edmonton, carrying about 9 tonnes per flight [139], and from there flown to Eldorado's refinery at Port Hope, Ontario [12] (see also Figure 5.9).

In the early years incoming aircraft could only operate in the winter, when they could land near the mine on the ice covering Lake Athabasca. The Gunnar airstrip was constructed in a wide, shallow valley close to the mine site, and opened in 1954 (Figure 3.14), with the first landing at the airstrip occurring on October 5 of that year [118]. From that date forward the airstrip enabled year-round air services[32]. This was all in place by the time production had commenced in 1955 [150]. *The Northern Miner* reported that Gunnar-Nesbitt Aviation carried 3,326 passengers and 1,039,139 kg (2,290,909 lb) of freight in 1956 [146].

In addition to flying yellowcake to Eldorado, the Gunnar aircraft flew daily from Edmonton, Alberta to Uranium City, Saskatchewan, providing free flights for the employees and their families [135]. The air service was also used to transport everything from construction materials and general supplies, to emergency parts, to groceries.

---

[32] According to Gilbert LaBine, the airstrip was only *"rarely closed in due to bad weather conditions"* [30].

GUNNAR URANIUM MINE

Figure 3.13. The Gunnar-Nesbitt DC-3C at the airstrip. Gunnar Mines Ltd.

Figure 3.14. 1955 photo of the Gunnar airstrip [118].

# 4 THE GUNNAR MINE

## 4.1 Establishment of the Mine.

The geology of the Gunnar deposit has been described in detail elsewhere [16,17,64,128,151-158]. Like the other main deposits of the Beaverlodge area, the Gunnar uranium deposit occurred as a vein in the Tazin formation of sedimentary and granitized rocks (granite gneiss and syenite) [10]. The granite gneiss and syenite range in colour from pink to mauve or grey, and consist of medium- to coarse-grained quartz and feldspar with minor amounts of chlorite and mica [10,64].

The Gunnar ore body was a predominantly pitchblende vein in an altered granite [154]. The age of the pitchblends has been estimated at about 750 million years [157]. Similar to the other economic pitchblende deposits in the Beaverlodge area, Gunnar's occurred on a secondary fracture structure near (but not on) a major fault plane in in this case the St. Mary's Channel fault [15]. The Gunnar orebody extended downward from the surface at a 45° angle [3,64,128,155]. The ore minerals were pitchblende and a small fraction of uranophane [3,10,153] and the ore is reported as being mostly fine- to medium-grained in texture and ranging in colour from pink to mauve to redish brown [64,153,155,157]. Other than the pitchblende and uranophane, the main gangue (i.e., unwanted) minerals in the ore-grade zones were calcite and hematite [3,10,128]. Drilling started in October, 1952 and showed such promising results that a large, systematic drilling program was launched almost immediately (Figures 4.1-4.3) [64].

About 190 drill holes were completed, totaling more than 21,000 m of drilling, in order to map the deposit ([17]). Ore-bearing rock was identified using a down-hole Geiger Counter and drill cuttings were collected and sampled for chemical analysis and ore grading (Figure 4.1)[33].

---

[33] An illustration of the processes of diamond drilling ore grading in this area and time period is given in a TMC documentary [57].

The main ore body was somewhat "pipe-like" in shape, with a diameter of about 140 m at the surface, angled at approximately 45°, and with a diameter of about 350 m at a depth of about 350 m (1150 ft) below the eighth level [3,151]. Most of the ore came from the upper ten levels (that is, down to a depth of about 440 m (1,450 ft) of the 13 levels that were explored [10].

The average ore grade was about 0.15% (as $U_3O_8$) in the host granite gneiss rock [3,16,17,151]. Most of the ore actually sent for milling was about 0.18 to 0.19% (as $U_3O_8$) [10,128,153].

**Figure 4.1. Diamond drilling at the surface. Gunnar Mines Ltd., 1963 [128].**

Figure 4.2. Blasting at the surface. Separate blast plumes from at least eight drill holes can be seen in this photograph. Gunnar Mines Ltd., 1958 [139].

Figure 4.3. Core samples from Gunnar. Author photo.

First, about 270,000 m³ of overburden had to be removed (Figure 4.4). This operation began in the Spring of 1954 [155]. The overburden comprised both permafrost muskeg and glacial silt, the latter being commonly referred-to as "rock flour" or "glacial flour" when it was dried [10,63,128,155]. The overburden thickness was as much as 20 m, and about 6 m on average [153,157]. This was removed with scrapers, a power shovel, and a dragline, and then the overburden materials were hauled away in "Euclid" trucks[34] [10,64]. In addition, over a million tonnes of surface waste rock had to be removed [10,63,155]. Originally (1954), this rock was first drilled using six wagon-mounted air-drills powered by mobile compressors, and then blasted [10]. By 1955 more sophisticated, self-propelled "Drillmaster" hammer drills with on-board compressors[35] were being used [10]. Most of the blast-hole drilling, whether for surface waste rock or ore, was completed with either 10 or 12 cm (4 or 4¾ inch) diameter tungsten-carbide drill bits [10,128]. For most of the blasting, drilled holes were loaded with 7.6 or 10 cm (3 or 4 inch) diameter by 40.6 cm (16 inch) long sticks of 75% Forcite[36] at the bottom, then sticks of 75% Dygel above the Forcite (to achieve continuous column loading), packed and covered with drill cuttings [128].

The waste rock or uranium ore were loaded using three 2.3 m³ (3 cubic-yard) Marion power shovels[37] into nine Euclid trucks with 11.3 m³ (14.8 cubic-yard) boxes [10,64] (Figure 4.5). Generally, one power shovel and one dump truck would be kept on standby while all of the others would be in continuous operation during each shift. Other mine and area heavy equipment included a "D-7" bulldozer, "Tournatractor" bulldozer[38], and a road-grader.

At first the blast-holes were tested at 1.5 m (5 ft) intervals with a Geiger-Müller probe, but with the advent of the Drillmaster drills (see footnote #35, above) it became possible to sample cuttings and dust from the drilling. The collected samples were given preliminary analysis by scintillation counter (which is more sensitive than a Geiger-Müller probe) and, selectively, more precise analysis by chemical assay [10]. These analyses

---

[34] Euclid Co. dominated the off-road, heavy hauling truck market in North America in the 1950s and 60s.

[35] These were pneumatic rock drills (also called drifters) that combined hammer drill and rotational motions to drill holes into rock while pushing the cuttings upwards to the surface for collection or disposal.

[36] Forcite was a "gelatin dynamite," 30 to 80% nitroglycerin mixed with cellulose, sodium or potassium nitrate, and a hydrocarbon like tar (to make it waterproof). Dygel was a "semi-gelatin" dynamite formulation.

[37] One shovel-full contained about 5 tonnes of rock or ore [139].

[38] Tournatractors were bulldozers with large rubber tires rather than tracks, enabling them to operate at greater speeds.

were not only used to distinguish ore from waste, and monitor ore grade, but also for ore-blending to ensure a constant grade could be fed to the mill. Exploration drilling continued throughout most of the operating lifetime of the mine. By 1961, just over 200,000 metres (67,700 ft) of core had been drilled and examined and/or analyzed [128].

Figure 4.4. Overburden removal, *circa*. 1954. Courtesy of Saskatchewan Archives Board (Photo RB5401-15).

The Gunnar mine and mill operated year-round, so equipment and lubricants had to be adjusted and improved upon from what was conventionally available at the time, usually by trial-and-error testing and providing operational feedback to the company's suppliers. In many cases the equipment had to be kept running, so it was self-heated, to reduce down-time when temperatures fell below -29 °C. The power shovels, for example, were continuously operated from mid-October through to mid-April of each year and only shut-down when necessary for oil changes or repairs [128]. Similarly, cold weather and vibration caused repetitive breakages of lubrication and hydraulic lines, as well as operating lights. Further details on their cold weather experiences are provided in references [10,64,128].

Figure 4.5. Loading uranium ore for transportation to the mill. Gunnar Mines Ltd., 1958 [139].

The surface waste rock was mostly used in the construction of the benches and road in the open pit, but about 15% of it was used to make foundations for the mine and mill buildings [10,128].

In the Fall of 1955, with mining from the open pit well underway, the construction ("sinking") of the underground shaft was begun [159]. A temporary wooden headframe was constructed in October, 1955, with the permanent steel headframe planned for the Fall of 1956. The first phase of the underground shaft was completed in the summer of 1956 and then held until the permanent headframe was completed in January 1957 [127,160].

In Section 1.4 above, it was noted that the health effects of low doses of radiation were not yet well known in the 1950s, so uranium mining was still considered "safe" (as far as radiation hazards were concerned). In the television documentary film, "*The Road to Uranium*" [59], a reporter speaks to one of the underground uranium miners at the nearby Eldorado mine in 1957, and asks "*Is uranium dangerous to work with? Do you have to be careful?*" to which the miner replies "*No ... [I] never experienced anything anyway.*"

## 4.2 Mining.

***Open Pit Mine.*** The Gunnar mine had the only open pit operation of any of Canada's Atomic Age or Cold War Era uranium mines [3] (see Figures 4.6 - 4.8). The open pit mine was roughly oval in shape, about 305 m by 244 m, and was mined to a final depth of 110 m below the level of the lake [3]. The rock was mostly mined in 9 m vertical benches with approximately 6.4 m berms on each bench [3]. The roads for trucks hauling ore and waste rock extended from a single entrance at the top of the pit, downwards at an 8% grade and 15 m width [3,64]. By the time the open pit had been mined to its final size, the distance from the bottom of the pit, along the roads, to the mill crusher was about 1.8 km [3].

The first sales contract for Gunnar uranium was signed in October, 1953, covering nearly 3,700 tonnes of $U_3O_8$ to be delivered to Eldorado between 1955 and 1960 [12]. Limited production began in September 1955, and the open-pit mine and mill officially began production on October 25, 1955 [161]. The mine was declared to be in full commercial production on March 1, 1956 [146]. At the time it was the world's largest and highest-grade open-pit uranium mine, keeping Gunnar in the news [162,163]. It was estimated that the Gunnar reserves would last for 12 years [161] although, even with subsequent reserve additions, the mine would ultimately only operate for another eight years.

At first, most of the production was waste rock. In 1957, Gunnar mined 1.762 million tonnes from the open pit mine, of which 518 thousand tonnes were ore and 1.244 million tonnes were waste rock [128]. However, as mining progressed the fraction of waste rock decreased, especially once the underground mine was opened. By October 1958 the ore to waste rock ratio had greatly improved, to about 1:1 [164].

The open pit was located very close to the shores of Lake Athabasca, separated by a bedrock ridge that is only about 90m wide. Despite the close proximity there was little subsurface flow from the lake to the pit or to the underground workings of the mine. During 1963, when underground development below the pit bottom essentially reached the maximum depth, pumping from underground averaged only 340 litres/minute (75 gallons/minute), which included the ingress of water from backfilling with tailings [128]. The ratio of waste rock to ore from the open pit mine was originally about 2.4 to 1 [10], falling to about 1.5 to 1 by 1957, and averaging 1.83:1 between 1957 and 1961 [128]. During winter operations, ore had to be moved to the mill as soon as possible after blasting otherwise, if allowed to cool and crushed when cold, it tended to adhere to conveyor belts and ore bins [15]. The open pit mine was officially closed on October 31, 1961, after producing 2,651,708 tonnes (2,923,603 tons) of ore [130].

The underground mine was developed slowly to provide a transition as the open pit began to be exhausted. The underground mine and hoist system went into operation in May 1957, supplied about half of the production in 1959, and completely superseded open pit production late in 1961, after which the mill was only fed from the underground workings [3,10,129].

Figure 4.6. (a, Upper). Opening up the open pit mine *circa*. 1955. Source: Saskatchewan Research Council; (b, Lower) Loading waste rock in the open pit. Gunnar Mines Ltd. 1963 [128].

Figure 4.7. The Gunnar open pit mine in 1959 as seen from the inside (upper) and from above (lower) [129].

**Figure 4.8. The Gunnar open pit mine, headframe, and mill circa 1963. From reference [128].**

The underground mine shaft was adjacent to the open pit mine and had three compartments: a 3.6 m by 1.8 m cage compartment, a 1.7 m by 1.8 m skip compartment, and a 1.7 m by 1.8 m personnel access, utility, and maintenance compartment [64,128]. In the first and largest compartment a mine cage, suspended from a hoist on steel wire rope, was operated like an elevator to transport miners and equipment. In the second compartment a skip, also suspended from a hoist on steel wire rope, was operated like an elevator to transport mined ore and waste rock up to the surface. The third compartment was used primarily for utility access (air, water, electrical, and ventilation) and maintenance, but was also intended to provide an emergency exit (see Figure 4.9).

It appears that there were at least two ventilation "raises" separate from the ventilation provided by the main shaft, one raise to the bottom of the open pit, and another raise surfacing between the mill and the open pit, near the powerhouse [134].

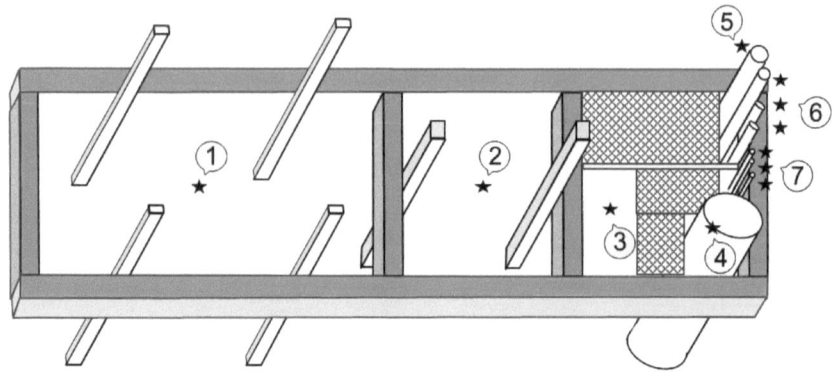

Figure 4.9. Illustration of the underground mine shaft compartments, showing the cage compartment (1), skip compartment (2), emergency exit (3), ventilation (4), high pressure air (5), water supply and drains (6), and electrical conduits (7). Adapted from information in reference [128].

Above the mine shaft was built a "permanent" 48 m tall structural steel headframe[39] that also housed a shaft house and two 907 tonne storage bins. A double-drum, 15,880 kg (35,000 pound) maximum-pull electric hoist was driven by two 400 hp, 400 rpm DC motors, which were in turn powered by two 350 kW diesel generators [64,128]. The hoist ropes were 3.8 cm (1.5 inch) diameter, 6 x 27 construction, Lang's Lay steel hoisting rope[40]. This combination was capable of hoisting either the cage or the skip at a rate of 8.1 m/s (1,600 ft/min) from the full depth of the underground mine [128].

During mining operations, the hoist would raise the skip to the surface and discharge the mined rock ("muck") into one of two discharge chutes – one chute for ore and another for waste rock. The chutes led to 907 tonne (1,000 ton) bins, an ore bin and a waste bin respectively. The Euclid trucks were driven under the storage bins from which they were directly filled with either ore or waste rock, by gravity feed.

Further details on the headframe and hoist operations are provided in references [10,64,128].

*Underground Mine*. The underground mining method used is called

---

[39] Decades later, the headframe eventually became dangerously unsafe and SRC had to demolish it, in 2011. See Section 8.5.
[40] This kind of wire rope comprised six multi-wired strands, with 27 wires per strand, "laid" or helically bent around a core. Lang's Lay refers to the way wires are placed within each strand, in this case meaning that wires in the strands were laid in the same direction as the "lay" of the strands around the core.

"*long-hole open stoping*" [3,128]. Stoping means that the ore was removed from spaces or passageways called drifts, which were then left open (Figures 4.10 and 4.11). This method was possible because the surrounding granite was stable enough not to collapse after the ore had been mined out. Long-hole stoping refers to the practice of drilling a pattern of holes, filling the holes with explosive, and then blasting to break-up the ore[41].

Figure 4.10. Drilling oversize rock in the underground mine. Gunnar Mines Ltd. 1963 [128].

---

[41] An illustration of the process of underground uranium mining in this area and period is given in a TMC documentary, using film footage from the nearby Rix Athabasca mine [57].

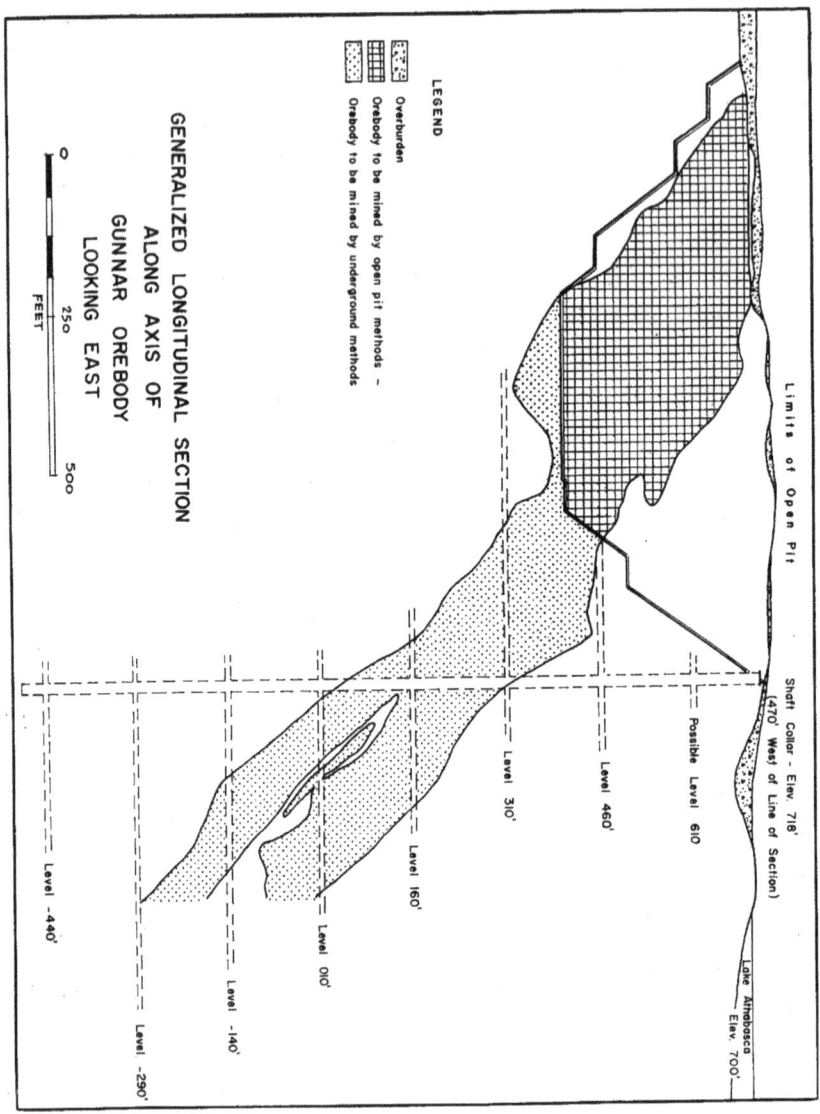

Figure 4.11. Illustration of the orebody showing the open pit, underground mine shaft, and the underground drift levels. Gunnar Mines Ltd., 1957 [64].

Since the orebody was aligned at about 45° from the horizontal, sections of steeply inclined ore called stopes would be identified for mining. At the bottom of a stope a haulage level would be established with ore recovery points (drawpoints) spaced at intervals along it (see the illustrations in Figure 4.11 and 4.12). From each drawpoint, in turn, a raise would be cleared to provide horizontal drifts to be built at intervals (sublevels) all the way to the top of the stope. From each drift, ore would be drilled with longholes, parallel to the haulage level and to right next to the face of the stope. From the ends of the longholes ore would be blasted free to fall to the base of the stope from which point it could be mined and transported along the haulage level (Figure 4.12). Some of the stopes thus developed contained more than 180,000 tonnes of ore and would typically comprise two or three drilling drift sublevels spaced about 27 m apart (vertically) [3]. The largest stope contained about 635,000 tonnes [128]. As each stope was mined out it was backfilled with waste rock and/or tailings in order to prevent subsidence *"since the orebody extended under the lake"* [3]. Blasted ore was collected and moved by rail to the main shaft for hoisting to the surface. For more details, see references [64,128].

The blasted ore was loaded into 2.6 m$^3$ (90 ft$^3$) "Granby-type" side-dumping ore cars connected into trains hauled by diesel locomotives to the skip-loading locations ("pockets") [140].

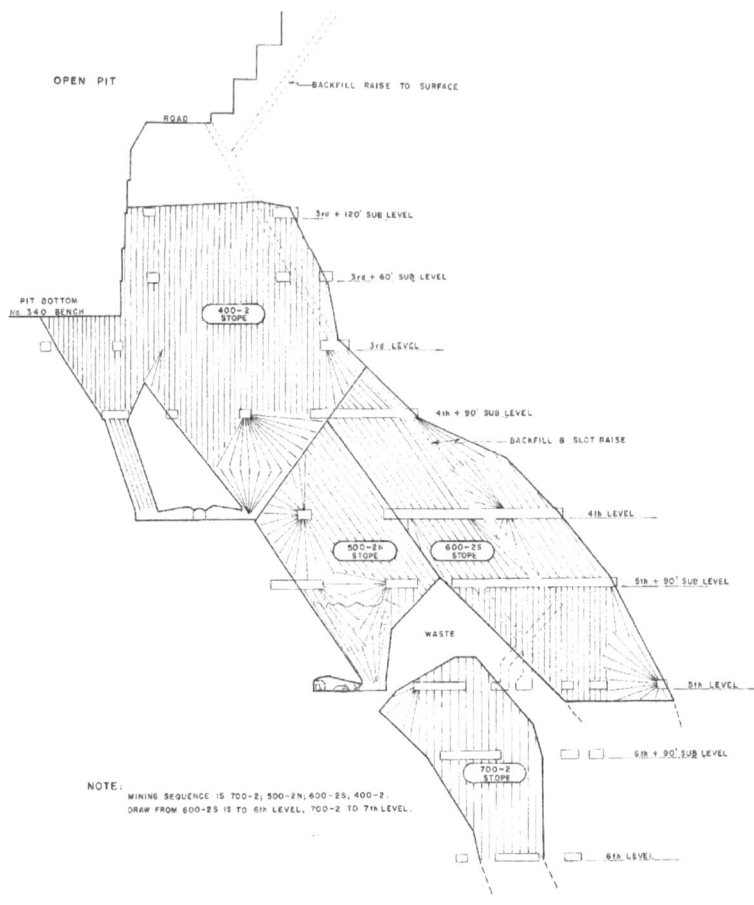

SECTION THROUGH UPPER MINING BLOCK

**Figure 4.12. Illustration of mining patterns in part of the underground mine. The planned stopes are shown; and the ore recovery points (drawpoints) would have been spaced at intervals along each marked level, where it intersects with the stopes. Gunnar Mines Ltd., *circa*. 1957.**

As underground mining progressed, water was encountered at about the 60 m (200 ft) level and thereafter water was semi-continuously pumped away. By 1957 the vertical mine shaft penetrated to 379 m, after which cross-cuts were driven to intersect the ore body, most of which lay beneath Lake Athabasca (Figure 4.11) [63]. Figure 4.13 shows a diagram of the above- and underground mines as of 1961 [165]. By the end of 1962 the shaft had been extended to 604 m, providing access to 13 discrete levels [10,128]. A computer reconstruction of the mines is shown in Appendix 3.

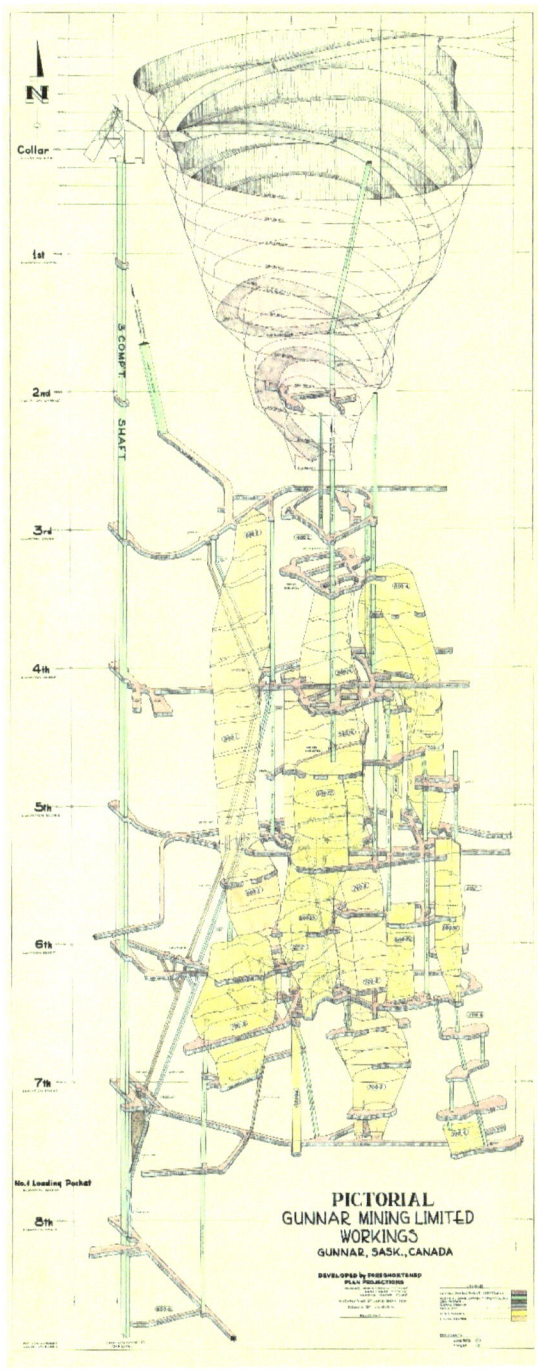

Figure 4.13. Diagram of the above- and underground mines as of 1961 [165].

In 1960 the company was able to start backfilling the first mined-out stope in the underground mine, using a combination of coarse waste rock and tailings [130].

In 1956 Gunnar's open pit mine produced 409,713 tonnes of ore and was considered to be the largest uranium producer in the world[42] [166]. In 1957 the underground mine also began production and the two mines together produced about 600,000 tonnes annually until 1963 (see Figure 4.14) [10]. During its main producing years, the Gunnar operation was also considered to be one of the lowest cost uranium producers in the world [167]. By the end of 1962 the Gunnar mines had produced about 4.7 million tonnes of ore [128].

In February of 1963 the stope pillar that separated the open pit from the underground mine was blasted open [168]. All proven reserves of ore were recovered from the underground mine and the last of the ore was hoisted to the surface on October 28, 1963 [168]. This final hoist marked the end of all mining operations at Gunnar, although milling operations continued into 1964 [168].

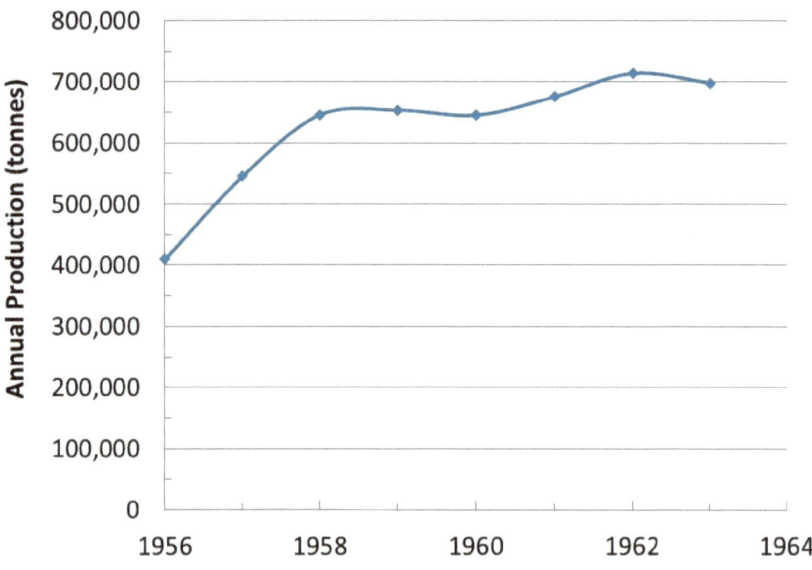

Figure 4.14. Annual production of uranium-bearing ore from the Gunnar mine including ore from the open pit and the underground mine. From data in reference [10].

---

[42] In later years Gunnar's yellowcake output was only eclipsed by the nearby the Ace-Fay-Verna mine.

## 4.3 Mine Infrastructure.

High-pressure (110 psi) air, ventilation air, mine water, and drinking water were all supplied from the surface, conducted down the "access" compartment of the main shaft, and distributed throughout the mine (see Figure 4.9). Ventilation air was particularly important as the flow rate had to be high enough to deal with radiation (from the uranium, radon, and their decay products), dust (from drilling and blasting), fumes (from diesel equipment), and blasting gases. Of these, the major hazards were considered to be radiation and dust, in that order, and these were semi-continuously measured and monitored [128].

A system of sumps on most levels, feeding pumps on three levels (levels 4, 8 and 12) was used to collect and remove water that constantly seeped into the mine (including drainage from the open pit mine) [63,64,128].

Ore and waste rock were transported underground by either air-driven or diesel trams depending on the amount of ventilation available in a particular location. The tram cars moved on a narrow-gauge railway having 24-inch gauge rails, on 12.7 cm (5") ties spaced 46 cm (18") apart [128].

A large equipment maintenance "service" group maintained the large number of heavy and light vehicles used in the mine and around the site (see Figure 4.15).

A dedicated mine engineering department managed such areas as construction engineering, mine planning, blasting engineering, pit engineering, surveying, design engineering, and mine contracting. Due to the remote location of the site, and the short seasons available for shipping by waterways a large, 929 m$^2$ (10,000 ft$^2$), warehousing facility was used to maintain stocks of spare parts and consumable goods, including oil, sulphur, and explosives [64].

Figure 4.15. 1955 photo of the mine equipment maintenance "service" building (background) [118].

# 5 MILLING

## 5.1 Background.

The Atomic Age and Cold War Era uranium mills were somewhat unique in Canada in that they were in very remote locations (making the shipping of process equipment and chemicals to the mills unusually expensive), with harsh winter climates (creating special problems of heating, moisture, and sometimes permafrost), they were required to efficiently process low grade ores (for which an alternative to the classical "gravity method" of separation had to be developed and then adapted to the particular ore from a given mine), and they required "EMF control" (to maintain a particular oxidation state on the part of the uranium being leached). The Canadian uranium ores contained pitchblende and uranophane with a large percentage of uranium (IV) [3], so the mills had to control the oxidizing nature of the leach solutions, which was done using sodium chlorate.

Most of these issues were dealt with in the development of the large mills before Gunnar, namely those at Port Radium and Beaverlodge [3]. The Canadian Department of Mines and Technical Surveys (part of what is now Natural Resources Canada) developed an alkaline leach process that was introduced at Port Radium in 1952, and then adapted for and introduced at Beaverlodge in 1953 [3].

In 1953 the Department of Mines and Technical Surveys began test-work with Gunnar core samples and then with the 54 tonne bulk sample mentioned in Chapter 2. They selected and further adapted an acid and sodium-chlorate leach process, complete with ion-exchange and subsequent magnesia-precipitation recovery processes that were ultimately adopted by Gunnar Mines [10,64,128]. Work also had to be done to adapt the conventional ion exchange process for recovering uranium from leach solution to produce a higher grade of uranium oxide (yellowcake) suitable for eventual processing into nuclear fuel [3]. The milling flowsheet design was finalized in December of 1953 enabling the equipment to be ordered at the beginning of 1954 [118]. Additional process optimization and troubleshooting was provided by the Saskatchewan Research Council (SRC), beginning in about 1958 [69,169].

Even the conventional methods for uranium assay were not completely suitable. C. Lapointe, of the Department of Mines and Technical Surveys, had developed a radiometric method for the analysis of uranium that was still in use for mine samples through to 1964 [3], but the uranium mining and milling companies operating in the Athabasca Basin in the 1950s, including Gunnar, needed a method that was faster so it could be used for process control in the mills [169]. Dr. Gene Smithson and others at (SRC) developed a new X-ray fluorescence-based analytical method for the determination of uranium in process samples that was both faster and more accurate than the earlier methods, and it was adopted in the Gunnar mill [169,170].

As noted above, the mill construction was completed and started-up in August 1955 with the first barrel of concentrate being filled the following month [10,128]. The mill reached its original design capacity (1250 t/d) by October of 1955 [171], and was expanded to 1650 t/d by March of 1957 [10]. The mill's average recovery efficiency was about 93% [3]. Over the almost nine years of operation, the plant milled 5.177 million tonnes of ore at an average rate of 1,677 tonne/day, producing 8,121 tonnes of $U_3O_8$ [3] (see also Chapter 7).

Gunnar's production was "second in tonnage only to the Ace-Fay-Verna mine" [154] (see also Table 1.1).

## 5.2 Mill Operations.

The Gunnar mill began production in September 1955 and was in continuous production through to February 1964 [3]. The milling process consisted of a number of operations, which have been described in some detail in the public literature [3,10,63,64,128].

***Crushing Circuit***. The raw ore was delivered from the mine by Euclid trucks and dumped directly into the first crusher (see Appendix 4). The ore was then crushed through a succession of gyratory and jaw crushers and screens that progressively reduced the largest sizes to 12.7, 3.8, and finally 1.6 cm (5", 1.5", and 5/8"). See Figures 5.1 and 5.2. Conveyors were used to transport the ore between crushers and grinders (see below), and to take crushed waste rock outside the mill. The process flow and layout are shown in Appendix 5. The crushing circuit was also used to crush about 100,000 tonne per year of waste rock to provide material for surface backfill, roads and road maintenance, and the construction of pads for buildings such as the school (Figure 5.3).

**Figure 5.1. Conveying crushed ore in the Gunnar mill. Gunnar Mines Ltd., 1958 [139].**

***Grinding Circuit***. Conveyors carried the finely crushed ore from storage bins to a rod mill in which it was ground to "21% plus 10 mesh" (that is, 79% less than 2.0 mm diameter) and split into equal streams feeding two spiral classifiers. From this point on there were two independent production lines. This arrangement enabled processing to continue in a single production line if the other one had to be shut down due to breakdowns, or for maintenance. The production lines next used ball mills to further grind the ore to produce slurries having 39 mass% solids of size "45% minus 400 mesh" (that is, 45% less than 0.037 mm diameter) – see Figure 5.4.

Appendix 6 shows the layout and process flow for most of the rest of the mill circuit.

**Figure 5.2. Outside view of the covered mill conveyor galleries. Gunnar Mines Ltd., 1958 [139].**

***Thickening Circuit.*** In each production line the 39 mass% slurries were thickened by sedimentation to 60 mass% in 7 m (23 ft) high thickening tanks. The thickener overflow was pumped back to the grinding circuits. The concentrated underflow was pumped to the leaching circuit.

***Leaching Circuit.*** In each production line the ore slurry was passed through a series of seven large 6 m diameter by 7.6 m high (20 ft by 25 ft) wooden-stave leaching tanks (Figure 5.5). In each tank the ore was exposed to air jets and sulphuric acid at a pH of 1.8. In each tank the pH was maintained by the addition of 93% sulphuric acid from the acid plant. A 25% solution of sodium chlorate was added to the first leaching tank as needed to maintain it in a slight excess condition, and therefore to maintain slightly oxidizing conditions at all times. In this way uranium (IV) was oxidized to uranium (VI), making it soluble in water.

**Figure 5.3. Conveying waste rock from the crusher in the Gunnar mill. This rock was used for various construction purposes around the site. Gunnar Mines Ltd., 1958 [139].**

The overall chemical reactions for uranium (IV) and uranium (VI) oxides in the pitchblende ore were:

$$UO_2 + 2H_2SO_4 \rightarrow UO_2SO_4 \ (aq) + SO_2 \ (aq) + H_2O \quad \text{for U(IV)}$$

$$UO_3 + H_2SO_4 \rightarrow UO_2SO_4 \ (aq) + H_2O \quad \text{for U(VI)}$$

The residence time for ore in the entire leaching circuit was about 16 hours [128].

Figure 5.4. Grinding circuit. Gunnar Mines Ltd., 1957 [64].

Figure 5.5. Gunnar mill leach tanks *circa*. 1955. Source: Saskatchewan Research Council.

***Filtration Circuit***. Filtration and then "clarification" (sedimentation) were used to separate undissolved solids from the dissolved uranyl sulfate. In each production line the ore slurry was pumped from the seventh leaching tank to a splitter mounted above four string-discharge drum filters[43] (Figure 5.6), at which point a filtration aid was added (0.25 mass% jaguar gum[44]). The filter cakes on the drums were washed with water, re-slurried to 57 mass% concentration, and flowed to a 4.9 m diameter by 4.9 m high (16 ft by 16 ft) wooden-stave surge tank. From the surge tank, the slurry was pumped to another splitter mounted above four secondary drum filters, and filtration aid was added again. The filter cake on the secondary drums was again re-slurried to 57 mass% concentration, and flowed to a tailings disposal tank. The primary and secondary filtrates were combined and pumped to the clarification circuit.

---

[43] These are drum filters for which a series of strings are drawn across the drum face to dislodge the filter cake.
[44] Jaguar gum is better known as guar gum, a cationic-polymer.

**Figure 5.6. A string-discharge drum filter in the Gunnar mill. Gunnar Mines Ltd., 1958 [139].**

*Clarification Circuit.* In each production line the ore slurry was pumped from the filtration circuit to a 9.1 m diameter by 6 m high (30 ft by 20 ft) wooden-stave thickener tank to which a 0.5 mass% solution of glue[45] was added as a coagulating/sedimentation agent. From this tank the sediment was withdrawn once per day and recycled to the leaching circuit. The overflow (supernatant) was passed through either a leaf-filter or a drum-filter (Figure 5.7), depending on the production line, and the filtrates pumped to 9.1 m diameter by 6 m high (30 ft by 20 ft) wooden-stave storage tanks in which the pH was maintained at 1.7. At this pH the uranyl sulfate was dissolved as $UO_2(SO_4)_3^{-4}$ *(aq)* ions. The solution was referred-to as "pregnant solution."

In 1959, a two-stage hydrocyclone plant and four hemispherical-bottom storage tanks were installed for the separation and storage of sand to be used for backfilling mined-out stopes (as described in Section 4.3 above) [3].

---

[45] This "glue" was probably an organic non-ionic polymer derived from collagen.

**Figure 5.7. Drum filters. Gunnar Mines Ltd.,** *circa.* **1957.**

*Ion Exchange Circuit*. An ion exchange process was used to recover a high concentration of uranium from the clarified acid-leach liquor. This ion exchange process had been demonstrated in South Africa but not previously used in Canada [172]. Each production line had four 2.1 m diameter by 4.3 m high (7 ft by 14 ft) ion exchange columns containing IRA-400[46] resin having chloride ions adsorbed at their exchange sites. In each production line the solution was passed down-flow (from top to bottom) of two of the ion exchange columns, in which the dissolved uranyl sulphate ions were exchanged for the chloride ions. The ion exchange reaction was:

$$UO_2(SO_4)_3^{-4} \ (aq) + 4RCl \rightarrow UO_2(SO_4)_3R_4 + 4Cl^-$$
$(R^- = \text{resin site})$

Most of the dissolved impurity ions passed through the columns, along with the released chloride ions, and were discharged to the tailings. When

---

[46] Amberlite IRA-400 is a quaternary ammonium "strong-base" type anion exchange resin that had been developed in the late 1940s.

the first column became saturated with uranium ions it would be taken offline, what was the second column would become the new "first" column, and the next column in line would become the new "second" column, and so on in repeating sequence. The uranium ions were then recovered from the columns by flushing them twice with a 1 molar solution of sodium chloride in dilute sulphuric acid, which would restore the column to the original chloride form. The ion exchange recovery reaction was:

$$4Cl^- \text{ (aq)} + UO_2(SO_4)_3R_4 \rightarrow UO_2(SO_4)_3^{-4} \text{ (aq)} + 4RCl$$
(NaCl to elute)

The effluent from the second elution would be used for the next column to be desaturated and the effluent from the first elution, containing concentrated uranyl sulphate was sent to the precipitation circuit.

***Precipitation Circuit.*** Each production line had two 4.9 m diameter by 6 m high (16 ft by 20 ft) agitated wooden-stave tanks in which the concentrated, dissolved uranium ions were precipitated with a 35 mass% magnesia slurry to form yellowcake. The precipitation reaction was:

$$MgO + UO_2SO_4 \rightarrow UO_3 \downarrow + MgSO_4$$

This process was monitored by pH and considered to be complete when the pH had reached 7.0 (which took about 8 hours). At near-neutral pH, and with oxidizing conditions no longer being maintained, some uranium (VI) would have become reduced to uranium (IV), yielding $UO_2$ precipitate as well.

***Product Filtration Circuit.*** Each production line had two plate-and-frame type filter presses that enabled batches of the yellowcake precipitate to be washed and filtered. The washing solution was treated with sodium chloride and sulphuric acid to create the solution used for column flushing in the ion exchange circuit above. The filter cake was treated with low-pressure air and placed into a hopper to feed the drying and packaging circuit. The solid uranium oxide concentrate is commonly known as "yellowcake." It has the approximate chemical formula $U_3O_8$, but is actually a mixture of uranium (IV) and uranium (VI) oxides and was approximately 2/5 $UO_2$ and approximately 3/5 $UO_3$. The final uranium precipitate was about 76% $U_3O_8$.

***Drying and Packaging Circuit.*** Each production line's hopper fed the yellowcake to a revolving-sweep batch dryer in which the moisture content was reduced to less than 6% in about 8 hours. At this point the two production lines merged again as a blower was used to lift the yellowcake to a single bag-filter storage hopper from which 113 litre (25 imperial gallon) steel drums were filled (Figure 5.8). These drums weighed about 204-227 kg (450-500 lb) each.

**Figure 5.8. Weighing a freshly filled barrel of yellowcake in the Gunnar mill. Gunnar Mines Ltd., 1958 [139].**

The drums of yellowcake were flown from the mine airstrip to Edmonton by either the Gunnar Aviation plane or an Eldorado Aviation aircraft (Figure 5.9), and from Edmonton the drums were shipped to Eldorado's refinery at Port Hope, Ontario [12].

All mining operations ended in October of 1963, and all milling operations ended in February of 1964 [168].

## 5.3 Mill Infrastructure.

Dedicated facilities were built for preparing the process chemicals, including laboratories, mixing areas, and a cold storage annex. These included dissolution and storage tanks, the latter being connected to pumps and piping to deliver the prepared chemical solutions to the appropriate circuits. The tanks were generally of wooden-stave construction with mixers (concrete and aluminum in the case of sodium chlorate).

**Figure 5.9. Loading a barrel of yellowcake into the Gunnar-Nesbitt DC-3C aircraft for shipment to Post Hope via Edmonton. Gunnar Mines Ltd., 1958 [139].**

An onsite laboratory with a staff of seven provided up to 140 chemical analyses/day [10,64]. Typical routine analyses included radiometric, fluorometric, and chemical analyses for uranium, plus "impurity analyses" for chloride, sulphate, iron, and phosphate [64,128].

For the leaching circuit sulphuric acid was made from elemental sulphur using a "contact type" Leonard-Monsanto process [173]. Sulphur from local pyrite was combined with sulphur imported from Jumping Pound, southern Alberta [63,113] (see Figure 5.10). An inventory of about 20,000 tonnes of sulphur (one year's worth) was kept in a stockpile near the acid plants [139]. The sulphur was melted and then used to feed two production lines, a 91 tonne per day (100 ton/day) line and a 59 tonne per day (65 ton/day) line (see Figure 5.11).

Figure 5.10. Unloading a barge of sulphur (for the acid plants) that had been transported from Alberta. Gunnar Mines Ltd., 1958 [139].

Figure 5.11. The acid plants in 1955 [118] and in 1958 [135]. Note the sulphur stockpile in the lower photo.

Sulphur was burned in air at slightly elevated pressure to form sulphur dioxide gas:

$$S(s) + O_2(g) \rightarrow SO_2(g)$$

The sulphur dioxide gas was then cooled and passed over vanadium pentoxide ($V_2O_5$) catalyst in a "converter" to form sulphur trioxide:

$$2SO_2(g) + O_2(g) \text{ (over } V_2O_5) \rightarrow 2SO_3(g)$$

The sulphur trioxide was then combined with sulphuric acid[47] in a countercurrent absorption tower to form disulphuric acid (also known as fuming sulphuric acid or oleum).

$$SO_3(g) + H_2SO_4(l) \rightarrow H_2S_2O_7(l)$$

The disulphuric acid was then combined with water to form very concentrated (~93%) sulphuric acid:

$$H_2S_2O_7(l) + H_2O(l) \rightarrow 2\ H_2SO_4(l)$$

The sulfuric acid was stored in two 1360 tonne (1,500 ton) capacity steel tanks. By-product heat from burning the sulphur was collected and used to make steam and hot water for the sulphur plant and other mine and mill facilities [173]. The sulphuric acid plant built in 1955 was capable of producing 91 tonne/day of what Gunnar called[48] "100%" sulphuric acid, and this was expanded to a capacity of 150 tonne/day in 1957 [3].

Mill operations were conducted with a staff of about 96 people working nominally 44 hour work weeks on three shifts, enabling 24 hour/day 7 day/week operations [10].

---

[47] Sulphur trioxide is not directly combined with water to form sulphuric acid because this is such an exothermic reaction that the product becomes acid vapour and/or mist rather than the liquid.

[48] Being made and stored without special precautions to prevent equilibration with the atmosphere it was actually 93% sulphuric acid, not 100%.

# 6 TAILINGS AND WASTE ROCK

Tailings from the milling processes were first collected in a 6 m (20 foot) diameter by 2.3 m (7 foot 6 inch) height wooden-stave tank, with a mixer the keep the solids suspended. A cyclone plant and associated storage tanks (added in 1958) were used to separate out sand for use as backfill in the underground mine. Otherwise the tailings were pumped to tailings ponds as a nominally 30 mass% slurry, through a 460 m (1500 foot) length, 25.4 cm (10 inch) diameter, redwood-stave pipeline that was supported by a wooden-trestle structure (see Figures 6.1 - 6.5).

Figure 6.1. The tailings pipeline is shown in the lower-left of this picture, *circa*. 1954. Courtesy of Saskatchewan Archives Board (Photo RB5401-20).

No tailings or effluent water treatments were applied to reduce contaminant loadings to the receiving environment [32]. The tailings were initially pumped into Mudford (Blair) Lake, now known as Gunnar Main Tailings (Figure 6.4). When the tailings reached the capacity of this area, a channel was blasted to allow the tailings to flow out to a local depression north of Mudford Lake, now known as Gunnar Central Tailings. Eventually the tailings filled this area as well, and overflowed into Langley Bay in Lake Athabasca [32]. This is clearly shown in Figure 6.5 below. The volume of tailings released was large enough to essentially cut Langley Bay into two distinct regions, one of which opens to Lake Athabasca and a smaller 'back bay' that is isolated from the lake except through a small channel [174,175] (Figures 6.5 and 6.6). Figure 6.6 shows the locations of the three tailings areas. This kind of tailings practice was common in the industry at the time, and disposal of tailings into nearby lakes was widely accepted as the most economical and convenient option [32].

Figure 6.2. The tailings pipeline and wooden trestle *circa*. 2005. Saskatchewan Research Council.

It has been estimated that 4.4 million tonnes of tailings were released into the environment [176,177] with a combined surface area of about 22 ha [178]. The surface areas have been estimated to be about 16 ha at Gunnar Main Tailings, about 4.3 ha at Gunnar Central Tailings, and about 0.7 ha in the depression between these two (see Figure 6.6) [178]. These areas do not include the portion of Langley Bay that contains Gunnar tailings. The combined tailings volume is estimated to be 130,000 m³ [179].

**Figure 6.3. The wood-stave tailings pipeline in 2006 (Author photo).**

The tailings depths have been estimated to be about 14 m at Gunnar Main Tailings, 3-4 m at Gunnar Central Tailings, and 2-4 m in Langley Bay [176]. In each case, the tailings are underlain by a peat or organic clay layer which is 0.5 to 9.4 m in thickness under the main tailings, 3 to 6 m under the central tailings, and 8 to 16 m under the Langley Bay tailings [177]. With an *in situ* permeability of approximately 10-7 cm/s, the underlying clay layer forms a reasonably tight physical and geochemical bottom seal. As a result, all of the water transported from the tailings occurs as either very shallow groundwater flow or as surface flows.

The overburden and waste rock was placed (mostly) in two large piles southeast of the mine pit (Figure 6.7). It has been estimated that 2.7 million m³ of overburden and waste rock were ultimately produced and placed into these two large piles, with a combined surface area of about 10 ha [174]. The waste rock is located on the shore of Lake Athabasca with the toe of one of the waste rock piles protruding into the water of the lake proper and into a shallow area immediately east of the waste rock pile itself [179].

Figure 6.4. Tailings as they are encroaching into Mudford (Blair) Lake, now known as Gunnar Main Tailings. Gunnar Mines Ltd., 1958 [139].

Figure 6.5. Aerial photograph showing all three tailings areas. The Gunnar mine site appears near the top of the photo. Base photograph courtesy Woodland Aerial Photography, 2004.

GUNNAR URANIUM MINE

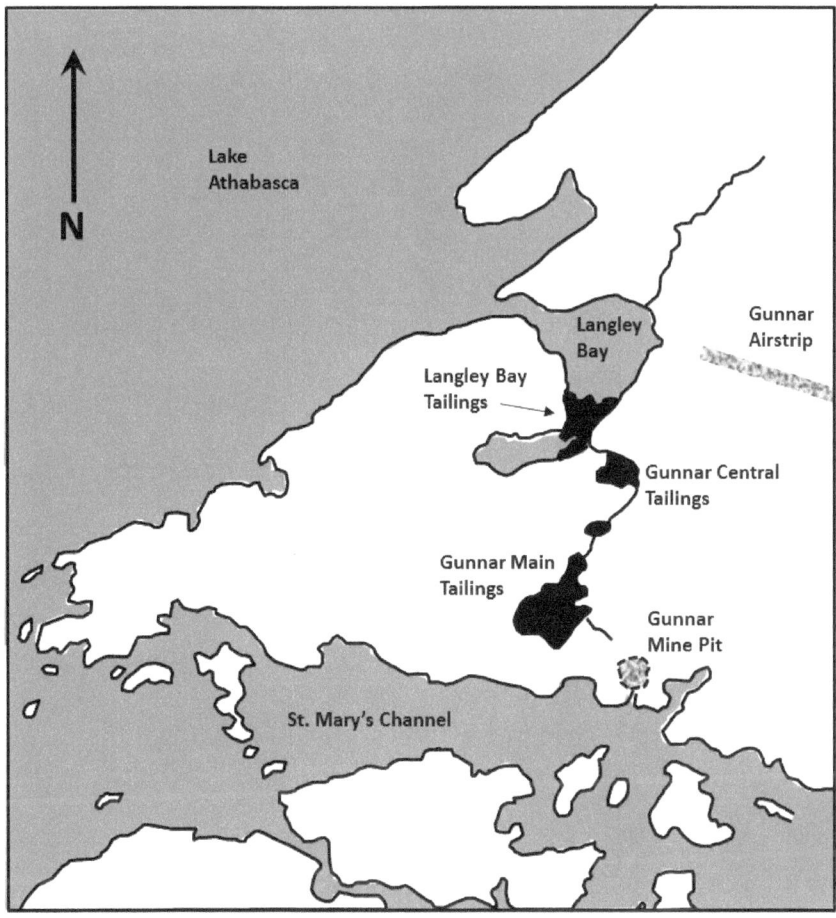

Figure 6.6. Illustration of the locations of the Gunnar tailings areas (shown in solid black). Map drawn based on Ruggles *et al.* [175] and Brown [183].

Figure 6.7. The main waste rock piles (foreground). Saskatchewan Research Council, *circa*. 2004 [180].

# 7 CLOSURE AND ABANDONMENT

The Uranium City boom that had begun in 1951 went through several ups and downs throughout the 1950s as the levels of prospecting and mine development in the area rose and fell. The established mining operations continued, however, and the company towns at Beaverlodge and Gunnar continued to support about half of the region's population [142]. At the end of 1961 Gunnar Mining reported reserves of 6,700,000 pounds (as uranium oxide) [132]. At the end of 1962 it was 3,450,000 lb (as uranium oxide) [181]. By the end of 1963 it was gone [168]. Also by 1963 the U.S. had more than enough uranium for its needs, the Rayrock, Beaverlodge, and Gunnar mines had all depleted their ores and closed and, although the Eldorado mill kept operating until 1967, the entire Uranium City area had fallen into decline [133].

## 7.1 Closure.

As noted above, all mining operations ended in October of 1963, and all milling operations ended in February of 1964 [168]. When the Gunnar site was closed it was essentially just shut-down without any decommissioning or reclamation [32]. This was not unusual at the time, as environmental protection was not yet a significant consideration for the industry or for the federal or provincial governments [32]. Accordingly, the Atomic Energy Control Board (AECB), which licensed the Gunnar mine and mill (and other nearby uranium mines of this era) had not imposed any decommissioning or remediation criteria requirements.

The company did, however, take some closure steps [168]:
- o upon the ceasing of mining in October of 1963, all underground equipment was removed, the mine shaft taken out of service, and

concrete bulkheads were installed on all surface openings to the underground mine,
- during the previous February the open mine pit and the underground mine had already been connected by blasting one stope pillar that separated the two, so that when the company blasted a channel between the open pit and Lake Athabasca, in December of 1963, the entire above- and below-ground mine workings were completely flooded up to the level of the lake itself (see Figure 7.1),
- the crushing and milling circuits were "cleaned-up" following the last mill production in February, 1964,
- the channel to the lake allowed the free movement of water (and presumably aquatic organisms) between the lake and the flooded pit until 1966 when the channel was blocked by filling it with waste rock.

Figure 7.1 Beginning the flooding of the Gunnar open pit in 1963. The channel that had been blasted between the open pit and Lake Athabasca is clearly visible. Saskatchewan Research Council [182].

The company had intended to sell, and/or reuse in other Gunnar Mining Ltd. operations the mine headframe, mine buildings, milling plant, acid plant, and equipment [168] but it is unclear how much of this actually happened. Subsequent Annual Reports of the company make little to no mention of the Gunnar mine, and do not record whether any significant sales of the remaining assets were ever made. It appears that some of the milling equipment was sold to Eldorado [183]. At some point in time the diesel power generators were salvaged [174,183].

The Gunnar town-site was abandoned. Many people moved from Gunnar to Uranium City and over the next several years many of the houses were moved by dredge over the winter ice road [135].

In August/September 1963 the Gunnar Water Transport Division's tugboat and all six of Gunnar's barges were dismantled at Fort McMurray and shipped by rail for reassembly and transfer to McNamara Corp. Ltd., a wholly-owned subsidiary of Gunnar Mining Ltd. [168].

The last official mention of the Gunnar-Nesbitt Aviation Company seems to have been in Gunnar Mining Ltd's. 1963 Annual Report [168]. It appears, however, that the Gunnar air services continued until 1965, at which time the Gunnar DC-3C aircraft (CF-GHX) was sold to Eastern Provincial Airways Ltd., and later to Gander Aviation Ltd. The nose and tail sections of this aircraft are on display at the North Atlantic Aviation Museum in Gander, Newfoundland (Figure 7.2) [149,150].

Figure 7.2. Tail section of the Gunnar DC-3C aircraft, in Eastern Provincial Airways colours, on display at the North Atlantic Aviation Museum in Gander, Newfoundland (Photo courtesy of the North Atlantic Aviation Museum).

The last Annual Report issued under the name Gunnar Mining Ltd. was in 1970 (for the year 1969) [184]. The company changed its name to Bovis Corporation Ltd. in 1971, and operated under the Bovis name through 1979. In 1979 Bovis appears to have amalgamated with several other companies to form the private company Kesmark Construction Ltd., which later the same year was renamed Kesmark Ltd. It appears that the shareholders of Gunnar Mining Ltd. and Bovis Corp. were paid out for their shares at this time, after which Gunnar and Bovis ceased to exist [185]. Kesmark appears to have ceased as a business somewhat later, in 1984.

Table 7.1. Final production statistics reported by Gunnar Mines [168].

| Open pit mine | | |
|---|---|---|
| Ore | 2,651,708 tonnes | 2,923,603 tons |
| Waste rock | 4,857,933 tonnes | 5,356,045 tons |
| Waste-to-ore ratio | 1.832:1 | 1.832:1 |
| Productivity | 106.6 tonnes/person-shift | 117.5 tons/person-shift |
| **Underground mine** | | |
| Ore | 2,620,766 tonnes | 2,889,488 tons |
| Waste rock | 321,019 tonnes | 353,935 tons |
| Waste-to-ore ratio | 0.122:1 | |
| Productivity | 24.7 tonnes/person-shift | 27.2 tons/person-shift |
| **Mill** | | |
| Ore milled | 5,175,929 dry tonnes | 5,706,647 dry tons |
| Average milling rate | 1,676 tonnes/day | 1,848 tons/day |
| Average grade milled | 1.69 kg $U_3O_8$/tonne | 3.38 lb $U_3O_8$/ton |
| Average recovery rate | 92.8% | 92.8% |
| Uranium produced | 8,121,118 kg $U_3O_8$ | 17,904,000 lb $U_3O_8$ |

## 7.2 Post-Closure.

The town was not totally uninhabited in the next two decades after the mine closed. For some years a caretaker remained onsite, and the town-site was partially inhabited for weeks to months at a time, between spring and fall, by teams of exploration geologists [135].

By the early1970s [135]:
*"The decay was a little more obvious in town. Weeds and small saplings pushed up along the sides of the road, and the street signs were starting to rust and tilt into the intersections. But the streets themselves were clear, the houses mostly intact, with only the odd smashed windowpane or swinging door. In the shopping centre, beams of light played through the skylight onto the dusty floors ... A row of counter stools with metal siding around the sides were lined up in front of the center of the cafeteria, the score of one last volleyball game still visible on the blackboard of the big gymnasium...."*

A fish processing and packaging operation was established[49] at Gunnar in 1971. Through the co-operative, the warehouse building and wharf were maintained, and possibly some other nearby buildings and/or residences [183]. Wash water and offal from the fish processing were dumped into the open pit during the fish plant's operating years [186]. The fish processing and packaging operations operated until 1980 with closure in 1981 [187]. The closure of the nearby Beaverlodge Mine, and the resulting reduction in barge traffic seem to have made it to too expensive to ship the fish south [135].

A visitor to the site in the late 1970s noted [135] that:

*"A street sign rattled in the breeze, waving the undergrowth back and forth; the windows of the houses gleamed in the sun like dormant eyes ... The mine still seemed eerily complete, like it had been abandoned just weeks before ... The town with its still intact buildings felt not just ghostly, but almost forbidden, like some terrible plague had wiped out its inhabitants and no one but us had dared to return since."*

In the years between closure and clean-up numerous scientific and other studies of the site were undertaken to understand the extent of the pollution and changes (positive or negative) occurring over time. These included studies related to the structures [174], open pit [176,186], tailings [175,176,178], waste rock [176], and the adjacent land and waterways [175,188].

When the Canadian Nuclear Safety Commission (CNSC) first visited the Gunnar site in 2000 [189] they found that:

*"... all of the structures were still standing ..., although some had been scavenged for useful parts. Equipment that had been shipped in by barge during operations still sat in crates near the docks. There was still yellowcake in barrels and hoppers in the barreling area of the mill. Sulfur piles were obvious around the acid plants and the unconfined uranium mill tailings had begun to migrate with the wind into the adjacent forest."*

In 2001 the CNSC reaffirmed that the site should be considered abandoned and under the care and control of the Province of Saskatchewan [190].

By the early 2000s many features of the original mine, mill, and town still remained in place on the site, including [174,195]:

o Utilidors,
o Headframe and shaft-house, including crates of equipment and engine parts,
o Mill buildings, including most of their internal piping and vessels

---

[49] Originally established by Athabasca Native Fisherman's Co-Operative in 1971, but taken-over by the Freshwater Fish Marketing Corporation in 1975.

(except the crushers, ball mills, and some leaching tanks, which had been removed), and including substantial quantities of leftover bulk chemicals,
- Other mine buildings, including the maintenance, warehouse, and office buildings,
- Acid plant buildings, including most of their internal piping, vessels, and tanks, and including large quantities of elemental sulphur and spent chemicals,
- Freshwater pumphouse, and nearby propane and oil tanks, and sewage treatment building,
- The dock and a derelict barge,
- Residence buildings, houses, and cabins (some of which may have been built after closure),
- The community centre, curling rink, and school.

**7.3 Hazards.**

In the 1950s it was generally believed that uranium mining and milling posed no particular radiation hazards as long as adequate ventilation was provided, although the need to control dusts was recognized [25]. At the time, both uranium mining companies and health authorities were aware that radium posed a health risk but information was lacking on specific exposure risks [25]. It wasn't until 1960 that Canadian regulations covering both uranium and radium exposure were brought in by AECB [25].

In the 1950s and 60s uranium mines in Saskatchewan were not subject to significant pollution control regulations and were allowed to use small adjacent lakes for tailings disposal, many of which tailings pond/lakes ultimately overflowed into other lakes [42]. As the 1960s unfolded, public awareness and concerns about radiation and acidity from abandoned and unremediated uranium mines in Canada began to emerge with regard to the Bancroft and Elliot Lake area mine abandonments [25]. The U.S. Government brought in regulations aimed at reducing uranium workers' exposure to radon in 1966 [23]. Nevertheless, at the time of Gunnar's closure, the closing-off of the open pit and underground workings by flooding was generally recognized as appropriate practice [183].

In subsequent decades the radiation hazards became much better understood, and public and government expectations for mine decommissioning in general changed substantially. As a result, expectations that abandoned mine buildings be dismantled and removed, and that tailings and waste rock be treated as necessary to protect humans and the environment were to emerge in later years.

Prior to SRC's remediation of the site several assessments were made of the physical, chemical, and radiological hazards, and of the risks these hazards presented to human and ecological health [33,134,174-180,183,186]. The principal hazards identified are summarized below.

***Structural Hazards.*** In a visit to the Gunnar site in 1993, Brown [183] noted *"... the very hazardous state of many of the buildings on site."* His observations were reinforced by the more detailed evaluation of KHS in 2002/2003 [174]. The 48 m tall headframe remained standing and covered an area 37 m by 23 m on a reinforced-concrete foundation wall pinned to bedrock. Although the headframe provided an obvious viewpoint over the lake, it had become unsafe with deteriorating stairway and cladding, and lack of guardrails. The mill and other mine buildings had become structurally unsafe, with deteriorating roofing, stairways, and cluttered access-ways. The acid plant buildings were in even worse shape, with heavily corroded piping, vessels, tanks, and access-ways. Most of the residences and community centre were visibly decaying to the point that entrances, stairways, and even some floors were unsafe. Leaking roofs had caused partial ceiling collapses in some of the structures. Even where structures had previously been demolished, most of the concrete floors and large amounts of structural concrete and steel remained in place.

Figures 7.3 and 7.4 provide some illustrations of the deteriorating buildings as they stood in the early 2000s.

***Radiation Hazards.*** Uranium mines and mills involve some unusual hazards in that they all have the potential for environmental impacts due to radiological toxicity. The half-lives of the radioactive contaminants of primary concern are:
- o  uranium-238 at $4.5 \times 10^9$ years,
- o  thorium-234 at 24 days
- o  thorium-230 at $7.5 \times 10^4$ years,
- o  radon-222 at 3.8 days, and
- o  radium-226 at $1.6 \times 10^3$ years.

In 2000 the Saskatchewan Government launched an assessment of its northern abandoned uranium mine sites in order to prioritize them based on public safety and environmental concerns [33], and the Gunnar site was rated as having the most severe such concerns [33]. Several other studies were conducted in the early 2000s [134,174].

The main radiation hazards to people were assessed as being [176,183]:
- o  Direct exposure to gamma radiation due to close proximity to radioactive materials in the tailings or waste rock piles,
- o  Inhalation of radon from the abandoned mine, mill, or tailings, or inhalation of radioactive dust blown from the tailings,
- o  Ingestion of radionuclides by drinking water or eating fish from Langley Bay or the flooded open pit.

Figure 7.3. Deteriorating residence buildings, in 2003. Saskatchewan Research Council [134].

Figure 7.4. Deteriorating leach tanks in the mill building, in 2003. Saskatchewan Research Council [134].

Scenarios involving such activities over a period of six months-time were projected to cause people to receive two- to four-times the annual dose limit for members of the general public, but well below the dose level set for atomic energy workers. Stated differently, it was estimated that the amount of time a person could spend in an area to exceed the public dose limit would be [174]:
- 8 days near the ore conveyor fines piles,
- 9 days near the Langley Bay tailings,
- 14 days near the Gunnar Main Tailings,
- 18 days near the Gunnar Central Tailings,
- 20 to 30 days in areas near the mill building, and
- 28 days near the waste rock.

A few areas in the mill building exhibited moderately high radiation levels ranging from 5 to 12 µSv/h (at 1 metre), notably in the string filter, product packaging, and ore conveyor areas [174,191]. Generally, other buildings around the site exhibited only low gamma radiation levels (0.1-1.5 µSv/h, at 1 metre) [174,191].

The tailings pipeline from the rear of the mill to the Gunnar Main tailings area showed evidence of tailings spills along its entire length, but the surface radiation levels were low, generally less than 2 µSv/h. Overall, the tailings areas exhibited moderately low surface radiation levels of about 3-5 µSv/h [144,174,179], but with some locations as high as 12 µSv/h [144]. The three tailings areas contain a significant amount of radionuclides, mostly following the uranium decay series of U, Th-230, Ra-226, Pb-210, and Po-210. Rain and snow melt can carry components of the tailings from Gunnar Main Tailings into Gunnar Central Tailings, and into the Langley Bay Tailings and Langley Bay itself, and therefore into Lake Athabasca [176,188]. However, chemical analyses showed that there are elevated levels of radionuclides and heavy metals in Langley Bay but not in Lake Athabasca [176,179]. This has been partly attributed to the presence of beaver dams controlling the flow out of the tailings area [179]. Similarly, elevated levels of radionuclides and heavy metals have been identified in vegetation growing on and near the three tailings areas, and elevated levels of radionuclides have been found in fish caught from Langley Bay [176,188].

The overburden and waste rock piles were found to have elevated concentrations of uranium and lead, and average radiation levels of about 1.5 µSv/h (with maxima of about 6 µSv/h) and average radon levels of about 250 pCi/L [144,174,177,179]. Regarding the potential for acid rock drainage and metal-leaching potential, the sulphur contents of the waste rock samples were found to be low (< 0.10 mass%). It was concluded that there is little acid generation potential for these materials [179]. Two small streams draining from the waste rock piles into Lake Athabasca were found to contain elevated levels of U and Pb-210, although the concentrations varied significantly by location, season, and even year [144,174,175,176,186].

The flooded open mine pit is meromictic, meaning that (the dense lower water layer is essentially isolated and does not mix with the upper layer). The pit has a surface area of about 7 hectares, a shoreline perimeter of 1700 m, and is about 110 m deep with a thermocline (temperature transition) at about 10 m and a chemocline (chemical composition transition) at about 75 to 90 m [186] – see Figure 7.5. In late fall of each year the thermocline largely disappears, allowing mixing from the surface down to about 55 m, while the chemocline remains intact throughout the year [186]. The concentrations of most heavy metals (except uranium) are low throughout the pit. However, uranium and radium-226 concentrations are quite high below the chemocline: about 2,800 µg/L for uranium [186]. Heavy metal concentrations are significantly higher in the pit sediment, especially for arsenic, chromium, copper, lead and zinc. There are populations of bacteria and phytoplankton in the pit. A 2002 study [134] found diverse and abundant communities of aquatic biota (phytoplankton, zooplankton, benthic macroinvertebrates, and macrophytes) and a healthy and self-sustaining population of northern pike (Esox lucius) and several other fish species [192]. Northern pike sampled from the pit in various years contained elevated concentrations of barium, mercury, selenium, arsenic, and radionuclides in their tissues and bones [176,186,193,194]. Fisheries and Oceans Canada determined that the Gunnar pit was not considered "Canadian Fisheries Waters" and therefore not subject to *The Fisheries Act* [192].

***Asbestos Hazards.*** Most of the mill and mine building roofs were covered with corrugated asbestos board. The exteriors of most, if not all residences had 'transite type" asbestos-shingle siding made of a cemented asbestos containing material (ACM)[50]. Inside many of the buildings and residences the walls had been sheeted with asbestos board, and many areas including ceilings, were insulated with spray-on asbestos which had then been painted-over. Asbestos was also found in some 'insulated' construction blocks. Generally, any building areas requiring extra insulation had been treated with spray-on asbestos to various thicknesses [174]. Asbestos insulation was also used extensively in the hot water heaters, boilers, and in the utilidors that connected almost all buildings and residences at the site. According to the Gunnar Mines company, Limpet

---

[50] Transite-type ACM products typically contained about 10-50% asbestos fibre, were common in the 1930s through 1980s, and were used in many applications in which a fire retardant was needed.

asbestos fibre was *"highly regarded ... for its insulating and fireproofing qualities"* [64]. As a result, and because it could be spray-applied, it was widely used throughout the site for insulation. In the post-closure era the asbestos insulation was generally in very poor condition and was found ubiquitously as litter in almost every structure.

**Figure 7.5. Aerial view of the flooded open pit at Gunnar in 2006 (Author Photo).**

***Other Chemical Hazards.*** Many of the mill and other mine buildings contained substantial quantities of chemicals, some in process vessels, some spilled in various locations, large quantities of spent process chemicals (including over 90 barrels of spent vanadium pentoxide catalyst pellets), and even many unused bulk chemicals. Examples of the latter were entire pallet-loads of bags of magnesium oxide and calcium hydroxide, some 15 pallets of Portland Cement, numerous barrels of sodium hydroxide, and large piles of elemental sulphur comprising several cubic metres [179] (see Figures 7.6 and 7.7). Fluorescent light fixtures were common around the entire site and some or all of their ballasts contained PCBs – the school alone was found to contain nearly 100 PCB-containing ballasts, the community centre had over 200, and another 200 were found in buildings around the site [195]. Some of the more than 8,000 barrels left in various locations around the site were filled, or partially-filled with unknown materials.

As noted above, numerous heavy metals and radionuclides are present in the flooded pit, waste rock, tailings and other areas. The metal Contaminants of Potential Concern (COPC) identified have been: antimony; arsenic; boron; cadmium; lead; manganese; molybdenum;

strontium; uranium and vanadium [179]. The uranium decay series of radionuclides (uranium-238, radium-226, thorium-230, polonium-210 and lead-210) have also been considered as COPC for their cancer-causing risk. Overall, the main COPC at Gunnar have been selenium, mercury, and uranium.

Figure 7.6. Example of pallet-loads of abandoned chemicals. Saskatchewan Research Council *circa*. 2006.

Figure 7.7. Illustration of the extent of the abandoned sulphur piles. Saskatchewan Research Council *circa*. 2010.

## 7.4 Status in the Early 2000s.

Very little had changed by the early 2000s, except for continued deterioration of the site. A very typical community perspective was offered in 2006 by Chief Victor Fern from nearby Fond du Lac [196]:

*"The buildings were wide open and there were no warning signs. The mine shaft was probably about 50 yards from the processing plant and that was wide open, too. We used to climb that every day. There was no fencing or anything. Even the uranium processing mill was wide open, so we used to go in all these buildings, not knowing the dangers about radiation. Also the tailings pond where all the waste went, it was fine sand that we didn't know was toxic waste, and we would go and play in there, not knowing."*

The Gunnar situation was not uncommon for uranium mines. Worldwide, only a handful of uranium mines have been completely or substantially remediated, including the original Shinkolobwe mine, Democratic Republic of the Congo, in the early 2000s [197]. According to the International Atomic Energy Agency, many parts of the world have experienced large delays in advancing the decommissioning and remediation of nuclear sites, and for a variety of reasons[51] [198]. Only a few of Canada's uranium mines have been completely or substantially remediated:

- o The Cluff Lake mine in Saskatchewan was remediated as of 2013 by AREVA Resources Canada Inc. [199],
- o The Agnew Lake mine in Ontario was remediated as of the early 1990s by Kerr Addison Mines [199],
- o The mines in the Bancroft area of Ontario (Dyno, Bicroft and Madawaska) were remediated in the 1980s and 1990s [199],
- o Of the 12 mines in the Elliot Lake - Blind River area of Ontario, five of the sites had been decommissioned by about 2002, and all of the rest have been decommissioned since that time. At the present time, all of these mine sites have been remediated, with their mine features capped or blocked, facility structures demolished, and the sites landscaped and revegetated [199,200],
- o The Rayrock mine in the Northwest Territories was remediated by the Government of Canada in 1996 [199],
- o The Port Radium mine, also in the Northwest Territories, was partially decommissioned in 1984 and fully remediated by the Government of Canada by 2009 [199,201],
- o The Gunnar and Lorado mines, plus some 35 smaller "satellite" mines in Northern Saskatchewan are currently being remediated by

---

[51] Including issues related to national policies and frameworks (or the absence thereof), financing, availability of technology and/or infrastructure, stakeholders, and/or politics – see reference [198].

the Saskatchewan Research Council (SRC) [202],
- The Beaverlodge mine in Saskatchewan comprises 62 licensed properties of which some have been fully remediated while others are still being remediated by Cameco Inc.

Orphaned or abandoned mines are of particular concern because they represent closed mines whose owner no longer exists, can't be located, or is unable to carry out remediation. In Canada, responsibility for the remediation of such mines reverts to government ("the Crown"). It has been estimated that there are about 10,000 such mine sites in Canada [203].

Canadian federal regulations did not cover uranium mine closures or remediation until May 31, 2000 when the Nuclear Safety and Control Act (NSCA) replaced the Atomic Energy Control Act (AECA). The new legislation was constructed to regulate the complete life-cycle of nuclear activities. As a result sites like Gunnar that previously existed outside of the jurisdiction of the AECA did come under the jurisdiction of the NSCA and meant that the Gunnar site would have to be licensed by CNSC [189,190]. CNSC created a Contaminated Lands Evaluation and Assessment Network Program to identify such sites, evaluate them for safety, and make recommendations for the regulatory approach to each site. As part of this process CNSC issued a temporary exemption from the requirements of the NSCA to enable the Province of Saskatchewan[52] to possess, manage, and store nuclear substances at the Gunnar site [189]. This enabled the province to take steps to secure the site and make plans for the ultimate remediation of Gunnar[53].

In 2003 the Government of Saskatchewan took a number of steps to limit public access to the main mill and mine buildings including [174]:
- Chain-link fencing was installed to limit access to the capped mineshaft,
- Existing doorways were sealed shut (usually by welding), and open doorways, accessible windows, and other access points were secured with chain-link fencing permanently attached (usually by welding),
- Many interior stairways, and all exterior ladders, were removed to a height of 2.5 m from the ground, and
- Numerous highly visibility warning signs (in English and Dene)

---

[52] It was accepted at this time that the Gunnar mine and mill site were considered abandoned and that its care and control had reverted to the Province of Saskatchewan.
[53] As one of these next steps, the province initiated discussions with the Government of Canada seeking funding support given that the establishment of Gunnar and all of its operations and products were controlled by the federal government and its strategic national defense and foreign policy interests.

were installed throughout the site that specifically identified the dangers due to radiation, asbestos, and failing structural integrity of the buildings, and warning the public not to enter.

Studies on channels and bays near the Gunnar site were conducted in 2004 and 2005 [193]. Some of these areas are contaminated by tailings, while others have received waste rock pile seep contamination. The uranium concentration in waste rock seep from the toe of the waste rock pile is high and may cause potential adverse effects on aquatic species in the wetland area into which the seep flows. For example, there are elevated radionuclide levels in the sediment near the channel that previously connected Gunnar pit to Lake Athabasca, and higher uranium levels in the fish tissues when compared to the fish from the reference area. By this time, the contaminated bay and channel areas were generally being re-colonized by a diversity of aquatic vegetation which provides habitat for fish and a food source for wildlife [179]. A number of the fish analyzed continued to demonstrate elevated radionuclide levels compared to the reference fish. The assessment of exposure to terrestrial wildlife to radionuclides indicated that there are no risks of adverse effects from radiation exposure. Exposure to non-radionuclides showed that uranium is an issue for terrestrial animals with a large aquatic diet such as beaver, ducks, mink and muskrat. Uranium concentrations in aquatic plants, benthic organisms and sediments are the main contributors. Radionuclide assessments [204] showed that, in general, releases from the Gunnar site did not pose any risk of adverse effects to aquatic biota with the exception of aquatic plants in nearby bays and the area close to the waste rock seep, for which radionuclide concentrations were substantially higher than background.

**7.5 The Iconic Gunnar Headframe.**

Possibly the most controversial of the demolition activities was the demolition of the headframe (Figure 7.8). As noted above, after closing the mine Gunnar Mining Ltd. had originally intended to sell the mine headframe or, failing that, to relocate it to other of the company's operations. This obviously did not happen, and the company's Annual Reports do not provide any explanation. Probably the cost of dismantling and shipping the headframe exceeded its market value.

Given that it was ultimately left in place, the headframe was the single most visible feature of the site (other than the water tower) and it became not only a landmark and aid to navigation, but also a symbol of the site and its history.

Demolishing the buildings at Gunnar, particularly the mine's iconic headframe was positive in terms of human health and safety but a sad development for many local residents, who saw it as a loss of a sign of their history:

*"To me, I don't want to see it go, because it's kind of like one of our main attractions. It's kind of our little museum up here."* (Long-time local resident Allen Augier, quoted in reference [205].)

For many people that lived-at, or have been associated with the Gunnar mine the principal image they remember from this period in time is the headframe at *"the Gunnar minehead, looming dramatically over the sweep of Lake Athabasca"* [135]. Although there was local community interest in establishing some kind of historical preserve, and possibly even a tourist attraction at the Gunnar site, the hazardous condition of the standing structures (including the headframe) made preserving them impractical, and to date lack of funding has precluded the construction of a museum or interpretive site.

**Figure 7.8. The iconic Gunnar headframe looms above the maintenance buildings in 1958 [139].**

# 8 REMEDIATION: THE ENDING

The Gunnar site illustrates the wide range of adverse impacts of abandoned, orphaned mines, which can include:
- o altered landscape, introducing hazards and the loss of otherwise productive land,
- o altered vegetation, and the loss, or at least reduced quality of vegetation,
- o altered groundwater, introducing hazards and the loss of otherwise potentially productive water,
- o altered water bodies and their sediments, introducing hazards and the loss, or at least reduced quality, of the water and/or its constituent plants and animals (such as fish),
- o air pollution from dust and/or hazardous gases,
- o physical hazards, including standing structures, pits and shafts, chemical piles and/or tailings dumps.

In cases where abandoned mines have left large amounts of tailings and/or waste rock deposited into unsuitable places then the sheer volume and mass of material involved mean that it is usually cost-prohibitive to move them to a more suitable location. In such cases the only practical option may be to allow them to remain in place and conducting some remediation to minimize ongoing harm to the environment [183]. These issues were also applicable to the Gunnar site.

The Gunnar site's visibility, mostly due to its size and the tall headframe near the lakeshore, the numerous hazards around the site, and its proximity to local communities led to various calls for clean-up and remediation over the years [183,196,206]. The Gunnar site is easily accessed by boat and was a popular picnic and hunting and fishing site, with its historic ghost town, mine structures, and the iconic Gunnar headframe.

The Gunnar site is immediately across the lake from the Athabasca sand dunes, which were becoming a recreation area and even somewhat of a tourist attraction.

## 8.1 The Remediation Process Begins.

In 2006, the Governments of Saskatchewan and Canada signed a Memorandum of Agreement (MOA)[54] to proceed with the decommissioning and reclamation of the Cold War legacy uranium mine and mill sites in Northern Saskatchewan [207]. The MOA included the remediation of the Gunnar mine and mill site. As the property owner, the Government of Saskatchewan had primary operational and legal responsibility for the project. The Saskatchewan Research Council (SRC), a provincial Crown corporation, was contracted as project manager and designated agent to manage and perform the required environmental

---

[54] Some of the historical background to the MOA can be found in reference [206].

assessment requirements and rehabilitation activities [207]. This included SRC acting as the "proponent" with regard to licensing by the Canadian Nuclear Safety Commission (CNSC). Generally, subject to regulatory approvals, it was anticipated that the physical project would consist of [144,179]:
- o Demolition of existing building, facilities and structures,
- o Appropriate disposal of materials resulting from demolition,
- o Installation of an appropriate cover on all or a portion of the exposed mill tailings as required,
- o Rehabilitation of the existing waste rock piles as required,
- o Rehabilitation of additional risk(s) as warranted,
- o General site clean-up,
- o Re-vegetation of areas of the rehabilitated site as required, and
- o Appropriate monitoring during and after rehabilitation.

The Canadian environmental regulatory regime is complex with both the provincial and federal government legislative frameworks applying. The provincial government is the owner and manager of the Crown Land on which the mine is located and designated permits are required to conduct any work on Crown Land. The provincial government also has approval requirements under provincial environmental assessment (EA) and environmental protection legislation. The provincial government has a joint agreement with the federal government that allows a coordinated provincial/federal approach to environmental assessment. The federal government has authority under federal environmental assessment legislation, fisheries legislation, navigable waters legislation and environmental legislation. The responsible federal departments' oversight is coordinated through the Canadian Environmental Assessment Agency but each department has distinct regulatory applications and authorities. In the uranium industry the federal regulatory framework is made more complex with the CNSC being the ultimate regulatory and licensing authority due to the presence of a "nuclear substance" onsite.

As an example, in 2007 SRC submitted [180] its project proposal for the rehabilitation of the Gunnar site for review and discussion with: Saskatchewan Environment, CNSC, Canadian Environmental Assessment Agency, Fisheries and Oceans Canada, Environment Canada, Natural Resources Canada, Transport Canada, Indian and Northern Affairs Canada, and Health Canada.

In an effort to ensure coordination and a proactive approach, SRC formed a "regulators" committee, comprised of the various agencies listed above, which was used to discuss expected regulatory requirements, potential remediation approaches and their risks and expected regulatory controls [179,216].

## 8.2 Community Engagement.

Numerous local, regional and provincial stakeholders were and are interested in the Gunnar remediation plans, activities, and outcomes. These include the residents of the closest neighbouring communities of Uranium City and Camsell Portage (with a population of about 120), but also a much broader range of stakeholders. Given the Gunnar site's location in a remote northern area utilized by First Nations, Métis and Northern residents, communications with these local residents was of paramount importance and was a high priority for SRC throughout its remediation activities. All of the major short-term and long-term site-safetying and remedial activities proposed by SRC included consulting with the local population (see for example [134,208-210]).

SRC also commissioned a traditional knowledge and traditional land-use study so that the environmental assessment and subsequent remediation could be planned and undertaken in the context of traditional uses of the area. This study was conducted by the Prince Albert Grand Council. Similarly, a socio-economic assessment was conducted by SRC on the potential to use biochar as a soil amendment during land reclamation and revegetation, as a possible way to achieve mutually beneficial outcomes for northern communities as well as the remediation project [211].

A Project Review Committee (PRC) was formed in the early stages to provide a forum that would ensure involvement of each of the impacted communities and enable them to provide direct input on desired remediation endpoints and options as well as advice on opportunities to maximize the involvement of northern residents in the economic activities emanating from the project. The PRC was established with the assistance of the Prince Albert Grand Council (PAGC) and included elected officials from: Prince Albert Grand Council, Fond du Lac First Nation, Black Lake First Nation, Hatchet Lake First Nation, Settlement of Uranium City, Settlement of Camsell Portage, and Hamlet of Stony Rapids. The first meeting with local Chiefs and Mayors was held in conjunction with a broader town-hall meeting at the Ben McIntyre School in Uranium City in March of 2007, and guidelines for the PRC were finalized and signed by all parties at a meeting in Stoney Rapids in May of 2008.

Communications were also established early with the Northern Saskatchewan Environmental Quality Committee (NSEQC), which comprises representatives from the northern municipal and First Nation communities that are impacted by northern mining operations in the province, and in particular with the Athabasca Sub-Committee of the NSEQC. The NSEQC monitors uranium mining in Northern Saskatchewan to confirm environmental protection measures and ensure operations are conducted to increase the socio-economic benefits of the surrounding communities, and it also serves as a vehicle to enable northerners to learn more about uranium mining activities and to see first-hand the environmental protection measures being employed, and the socio-economic benefits being gained [212].

Beginning in 2005, SRC has held annual public meetings in Uranium City, including discussion of the activities being undertaken at the site and these meetings have included representatives of the NSEQC, the CNSC and Saskatchewan Environment. Other meetings have been held periodically in neighbouring Athabasca-basin communities. Tours of the Gunnar site were also provided periodically for members of the PRC and the NSEQC. This range of meetings provided many opportunities for community information exchanges, discussions, and feedback.

Some of the most common questions raised by local community members were:
- What are the impacts of the project?
- What are the remediation options?
- Are there any training opportunities?
- Are there job opportunities?
- How can we actively participate in the remediation?

Other initiatives have included communicating project plans and progress in accessible northern media such as radio, media interviews, flyers, posters, mail-outs, newspapers, and magazines. SRC developed and continues to maintain a *"Project CLEANS"* website ([www.saskcleans.ca](www.saskcleans.ca)) [202] for this purpose as well. Information has also been routinely provided to northern media for inclusion in their publications (e.g., *Opportunities North*). A key feature of much of these communications has been translating key project information into the Dene and Cree languages to ensure that the information being provided would be broadly accessible [192] (see, for example the warning sign in Figure 8.1).

These kinds of engagement activities were extremely well received by the communities and their leadership:

*"... SRC is including communities in the writing of the Environmental Impact Statement ... that type of inclusion has been missing in the past and it's a refreshing change to see ..."* Diane McDonald, Prince Albert Grand Council (2010, [213])

Overall, public support was very high for the project given that a mine site that had been abandoned for over 40 years was finally being cleaned up [144,179]. The consultation process helped ensure that the locally impacted community would be comfortable with the rehabilitation activities and final state of the mine sites.

Figure 8.1. The Gunnar head-frame in 2006 (Author photo).

Based on advice from the advisory committee and community meetings, SRC revised its procurement process for contract work on the project and also arranged training programs, all with the aim of promoting bidding from and hiring of northern contractors and northern employees.

One of the most significant challenges has been meeting local expectations for economic benefits given the limited project funds available. Efforts have been dedicated to training local communities and Aboriginal entities in such aspects as the tendering process, safety practices, equipment operation, and so on. The project work was compartmentalized to allow local participation in a variety of tasks including light equipment and tasks and tenders developed that encourage use of local workforces. Efforts were also made to allocate project funds to hands-on training opportunities for work occurring outside the tendering process.

When the project was approaching the 2010/2011 demolition phase described below, both the PRC and the communities were engaged in discussions of how to maximize local employment during this phase. Community liaison positions were created to coordinate employment and training opportunities for individuals at the community level. Programs were developed and put in place to train local communities and Aboriginal entities in the tendering process, safety practices, etc., to maximize the ability of local companies to bid on work and to maximize the number of northern residents qualified to work on the demolition.

For example, in 2011 a program was offered in the Athabasca Basin Region that was aimed at providing capacity building and employment opportunities. The program was cost-shared by the federal government's Aboriginal Skills and Employment Partnership, and involved the communities of Hatchet Lake First Nation, Wollaston Lake, Black Lake First Nation, Stony Rapids, Fond du Lac First Nation, Uranium City and Camsell Portage. The training curriculum included modules on: Construction Basics, Confined Space Entry Awareness, Transportation of Dangerous Goods (TDG), Workplace Hazardous Materials Information System (WHMIS), Personal Protective Equipment, Respiratory Protective Equipment, Asbestos Abatement and Awareness, Radiation Protection, First Aid, and CPR. Approximately 130 Athabasca-basin residents participated in this particular training program (not including the ongoing training that was done on-site as the demolition and other work was done).

As a result of initiatives like this, during the demolition phase (described below), 50% of the workers on site were Athabasca basin residents and, despite the fact that there were frequently over 60 employees per shift involved in the demolition of the buildings on site, there were no lost time injury incidents [192].

Key remaining goals are to ensure that at the end of the remediation of the site all stakeholders are satisfied that the site poses no significant danger to public health and safety, is not a source of ongoing pollution or instability, and allows for productive use of the land similar to its original (pre-mining era) uses, or at least for acceptable alternative uses.

## 8.3 Remediation Begins.

In April 2007 SRC submitted a letter of intent, and a project proposal and description, to remediate the Gunnar site under a CNSC license. As a result, CNSC extended the exemption from licensing until 2013 to allow an environmental assessment (EA) process to be conducted under the Canadian Environmental Assessment Act (CEAA).

As the work progressed and more information became available the understanding of the various hazards at the site changed. By 2010 it was clear that many of the buildings that had been neglected for almost 50 years were structurally unsound and posed an immediate risk to people accessing the site [214]. Although the number of people on site was small, the situation prompted the CNSC to issue an Order under the NSCA (see Section 7.4) to conduct a safe demolition of all structures on the site by October 31, 2011.

SRC then developed a demolition plan [215] and implemented it, with some work being completed in the fall of 2010, notably the demolition of the wooden buildings on the site. Additional heavy equipment was brought in over a winter ice road (see Figure 8.2) and the rest of the demolition work was conducted during the 2011 field season, and completed on time by the fall of 2011 (see Section 8.5).

## 8.4 The SRC Camp at Gunnar.

Somewhat ironically, before the Gunner buildings could be demolished new infrastructure had to be built in order to provide on-site living accommodations, offices, material storage, fuel storage, and a laboratory (see Figure 8.3). The SRC camp ultimately included accommodations for up to 85 people, including kitchen, dining, and laundry facilities, plus independent power, water, and sewage treatment facilities. The laboratory and material storage facilities (a combination of trailer and "sea cans") were positioned near the camp and the laboratory is equipped with power and heat. Power is provided by diesel generators and/or battery storage, with (originally) 10 diesel storage tanks[55].

---

[55] One of the 10 tanks was subsequently demobilized from the Gunnar Site, with 9 tanks remaining on-site.

Figure 8.2. Transporting heavy demolition equipment over a winter ice road to the Gunnar site in spring, 2011 (Saskatchewan Research Council).

It was also necessary to make some improvements to the Gunner airstrip (widening and lengthening it to accommodate the aircraft expected to be using it, and to allow for a Medevac plane to land, if required), various roads around the site had to be improved to enable heavy equipment use, and a new docking area had to be created.

Figure 8.3. The SRC camp in spring, 2011 (Saskatchewan Research Council).

## 8.5 Demolition.

Numerous hazards had been identified at the mine, mill, town-site, tailings, and waste rock areas as summarized in Section 7.3 above. In the early 2000s, further assessments of the hazards at the Gunnar site were conducted and reviewed on a regular basis by SRC and the regulators [216]. It became apparent that one of the principal physical hazards to public safety on the site was the degraded nature of the standing structures and the potential for collapse, whether whole or in part, of many of the buildings and structures. Even for those that were deemed to not be at immediate risk of failure or collapse, most contained many internal hazards and were deemed unsafe for entry (e.g. missing floors, holes in floors, missing railings, degraded access platforms, etc.) [216]. It was decided in 2010 that virtually all of the standing structures on the site should be demolished without further delay, and a plan was developed for this to be accomplished by the fall of 2011. The demolition plan included dismantling of the various buildings and structures and temporary storage of the demolition debris, whose total volume was estimated to be 90,000 $m^3$ [216].

The general procedure for abatement and demolition of the buildings in 2010 and 2011 was to seal each area, remove and bag asbestos-containing materials and deposit them in temporary storage areas, remove and segregate other hazardous wastes, and demolish the building – usually involving a systematic dismantling of each structure piece by piece using powered equipment and excavators (see Figure 8.4) [216]. During this phase, numerous precautions had to be taken to minimize asbestos exposure to the demolition workers (see Figure 8.5). The various concrete foundations were usually just covered and left in place.

As discussed above in Section 7.5, probably the most controversial of the demolition activities was the demolition of the mine's headframe, which had stood as a landmark and historical symbol. Unfortunately, the headframe was found to be heavily corroded, structurally unsound, and covered in deteriorated friable asbestos cladding [195]. It was determined that it had to be demolished, and that the safest way to demolish the head frame was to remove the material that could be safely accessed, weaken the base, topple the structure, and then cut the steel beams into pieces for disposal. Explosives and guide wire were used to topple the head frame, which was then cut-up and removed to a storage cell (Figure 8.6). It also became necessary to re-cap the mine's main shaft (due to deterioration of the original cap) and also the ventilation shaft [195].

Overall, more than 80 buildings and structures were taken down in the 2010 and 2011 field seasons. Figure 8.7 illustrates the dramatic effect on the landscape of having demolished the buildings and structures at the Gunnar Mine site.

Figure 8.4. Dismantling of one of the residence buildings using a power excavator in 2010 (upper), and of one of the mill buildings using a hydraulic shear in 2011 (lower). Note the worker spraying with soap-amended water for asbestos abatement in the upper photo. Saskatchewan Research Council.

Figure 8.5. Collection and preparation of asbestos-containing and other hazardous materials for disposal. (Saskatchewan Research Council 2011).

Figure 8.6. Photographs of the headframe after detonation. (Saskatchewan Research Council, 2011).

Figure 8.7. Arial photographs taken before and after the demolition of buildings and structures at the Gunnar site. Courtesy Woodland Aerial Photography.

## 8.6 Post-Demolition Remediation.

*"Where are they going to store the stuff, even if they move it? ... and it's in the air, the water. It'll take a thousand years for this place to be safe."* (Former diamond driller William Schott quoted in reference [205].)

SRC completed an Environmental Impact Assessment (EIA) and filed its Environmental Impact Statement (EIS) in 2013 [192] (see Figure 8.8). The EIS provided a detailed description of the site, existing environmental risks, remediation approaches and their impacts [192]. It also provided recommended remediation options with mitigation measures, a suggested plan of action, and projected environmental outcomes both short- and long-term. Remediation options were considered based on managing ecological and human risks, as well as physical safety and human gamma-ray radiation exposures. The highest priority risk management needs were human gamma-ray radiation exposures and reductions in contaminant/radionuclide loadings from waste rock and tailings areas into areas frequented by fish.

Given that so much time had elapsed since the abandonment of the Gunnar Site very few records of operation were available and there remained significant uncertainties regarding some environmental aspects, so the advance selection of a single preferred remediation plan was not possible. Accordingly, the EIS used a series of decision trees to outline potential remediation steps and resultant impacts [192].

The decision-tree approach is based on the identification of potentially unacceptable risks to human and ecological health from gamma radiation, contaminants and physical structures. These risks were particularly evident at several areas including: all three tailings areas, the former acid plant area, the open pit, and the waste rock locations. Risks were analyzed in the immediate source area, as well as in the contaminant pathways and final receiving environment [216]. Potential remedial actions were sub-divided into:
- o those that are relatively straight-forward with little uncertainty, such as covering tailings, water diversions, removing contaminated materials from waste rock piles,
- o those that require additional monitoring data and follow-up assessment prior to selecting a final remediation approach, such as conducting further hydrogeological monitoring to determine contaminant sources and flow, and
- o those that are dependent upon selection of remedial approaches for other site areas that must be determined before the first one can proceed, for example, decisions on the flooded pit cannot be determined until it has been decided if the waste rock will be placed

in the pit or covered and left in place [192].

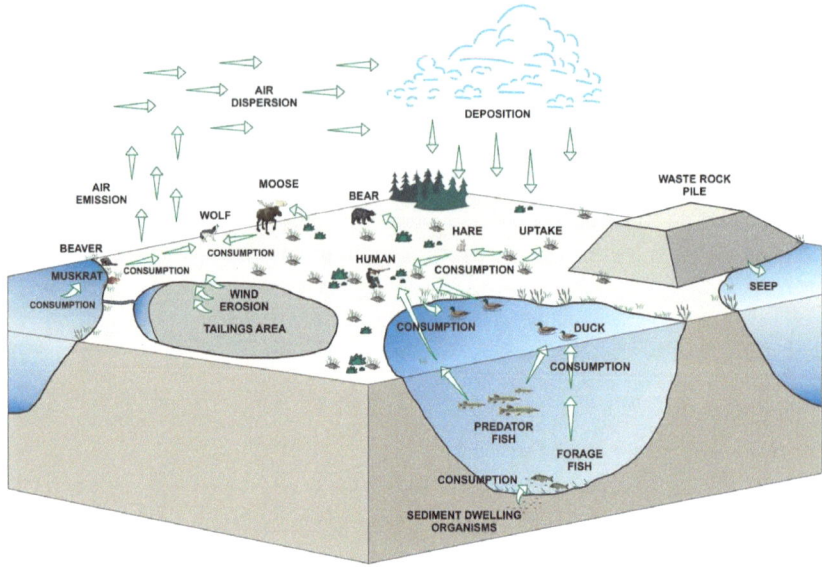

Figure 8.8. Illustration of some of the exposure pathways considered when conducting the environmental impact assessment. Saskatchewan Research Council, [182].

Decisions regarding the latter two categories will be made as part of the licensing phases. Even decisions in the first remedial action category will need to take into account all other decisions in order to avoid inefficiencies, such as multiple equipment staging and singular approaches to remediation versus a holistic approach [216].

Ultimately CNSC issued a 10-year Waste Nuclear Substance Licence to SRC in 2015 [217] (see Figure 8.9). The licence enabled SRC to proceed with the post-demolition aspects of the Gunnar remediation, but in staged phases, with regulatory "hold points" to enable recommendations and decisions about each stage to be made as the project progresses. The next major step was to deal with the tailings.

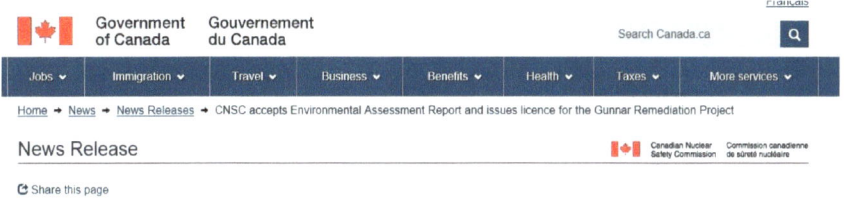

Figure 8.9. From a News Release announcing that a licence has been granted enabling the post-demolition phases of the remediation of the Gunnar site. Government of Canada, [217].

*Tailings*. The principal practical options for remediating the three tailings areas were natural attenuation, re-vegetation, rock cover and engineered cover [144,179,216]. Some of the areas had been re-vegetating naturally but with little change in the gamma-ray radiation levels, leaving a risk that the natural re-vegetation may provide a source of contaminants to local wildlife, windblown radioactive tailings dust still being an issue, plus the fact that some areas would never have re-vegetated naturally. The use of rock cover limits access to the tailings areas, reduces windblown activities and gamma-ray radiation, and can reduce water infiltration if it is underlain by a capillary-barrier layer.

Based on such factors as environmental impacts, technical feasibility, and cost effectiveness, the preferred remediation designs recommended by SRC for the three primary tailings deposits at the site were as follows [218]. For Gunnar Main Tailings, to completely cover the area, create a water-shedding landform by re-contouring with waste rock fill so as to direct surface waters to nearby Langley Bay, cover the re-contoured area with at least a 0.6 m thick layer of local till material, and revegetate the surface cover with native plant species. Similarly, for Gunnar Central Tailings, to create a water-shedding landform using waste rock fill so as to direct surface waters to an armoured drainage channel along the eastern perimeter, cover the re-contoured area with at least a minimum 0.6 m thick layer of local till material, and revegetate the surface cover with native plant species (see Figure 8.10). For the Langley Bay Tailings, to create a water-shedding landform using local till material or quarried fill that establishes a defined beach area based on the estimated high water level for Langley Bay, place large riprap[56] material along the east and south shorelines to protect

---

[56] An erosion-resistant ground cover comprised of fairly large, loose, angular stones.

the cover system from wave action and ice scour, construct an armoured drainage channel across the centre of the final landform cover, and revegetate the cover system surface with native plant species. The CNSC agreed with the recommended plan, and removed the hold-point related to remediation of the tailings areas in 2016 [219].

Figure 8.10. Illustrations of the before (upper) and after (lower) landform in the Langley Bay tailings area. Saskatchewan Research Council [218].

The tailings cover plan summarized above makes substantial use of Gunnar waste rock, which has the advantages of being a competent, coarse-textured material suitable for providing a working platform for construction equipment to place the final till cover system, and has a suitable particle size distribution for limiting the capillary rise of underlying tailings pore-waters into the cover system rooting zone. Also, being stockpiled waste this results in less disturbance of the natural landscape. This approach provides a financial advantage, in addition to the fact that the waste rock would have to eventually be moved anyway. As of the time of writing (Fall 2016), SRC was beginning to open up areas to provide clean fill (Figure 8.11) and laying out the routes for hauling materials in preparation to begin covering the tailings with the opening of the spring field season in 2017.

**Figure 8.11. Opening-up clean fill areas, in Fall 2016, to provide tailings-cover materials. Author photo.**

***Buildings, Support Facilities, Waste Rock, and the Flooded Pit.*** There are two main options[57] available for disposal of the demolished mine/mill buildings and all other structures: either to dispose the estimated

---

[57] SRC did consider shipping the steel to a Canadian or foreign steel producer for recycling, but these options were cost-prohibitive and suffered from significant public perception issues related to the transportation of radioactive materials.

80,000-90,000 m³ material in the mined-out pit (in which case the contaminated water presently in the pit would have to be managed while continuing to protect Lake Athabasca), or to landfill the material in one or more landfill sites (in which case suitable locations would have to be prepared and ultimately covered) [144,179,216]. A detailed hazardous materials inventory and a demolition strategy will have to be developed, as well as a site-specific health and safety plan.

The waste rock material had an original projection of low average gamma radiation, low acid generation potential and moderately elevated concentrations of uranium and lead. The most recent survey in 2009; however, has shown that there are substantial areas that exhibit higher than projected close out values. Options for remediation of the waste rock include: natural attenuation, contouring, engineered cover and re-locating part of the piles [144,179,216]. Re-locating the waste rock (any materials within 30 metres of Lake Athabasca) would require further contouring of the remaining slopes, significant operational safety hazards and disruption of the aquatic habitat. A more likely scenario could be the covering of any hotspots to ensure the average gamma-ray radiation levels are below 1 µSv/hr, plus minor stabilization where necessary, with minimal shoreline disturbance and no major re-sloping. This remains to be determined.

The flooded Gunnar Pit contains a total volume of approximately 2.6 million m³ (see the cross-sections illustrated in Figure 4.11 and Appendix 3). The channel connecting the pit to Lake Athabasca is filled with waste rock and this channel may have some hydraulic conductivity. Current studies show that the existing water quality of the pit does not pose significant risk to Lake Athabasca. The principal remediation options include: natural attenuation, safetying of the pit, sealing of the channel, and filling-in the pit. At the time of writing it is considered to be unlikely that the pit will be used as a disposal area [144,179,216].

One approach to stabilizing any of the current or future surface cover materials is revegetation, involving the use of plant species that facilitate a natural transition towards stable plant communities that reduce wind and water erosion and deep drainage of water. Rather than restoring the original vegetation, in some cases it may be better to introduce non-native species that are better suited to nutrient-lean, acidic soils. In the decades since Gunnar's closure several studies related to the potential for revegetation at Gunnar have been conducted [220].

## 8.7 Transfer to "Institutional Control."

It is SRC's intention to ultimately transfer the entire remediated Gunnar site into the Saskatchewan Government's Institutional Control Program (ICP), which falls under The Reclaimed Industrial Sites Act (2007), and for which the province has Reclaimed Industrial Sites Regulations [189]. The endpoint criteria for the Gunnar remediation project were developed with this process in mind, so the remediation endpoints had to be passive and require minimal maintenance over the very long term. The projected endpoints are [144,179,216]:

- radiation levels that do not exhibit in excess of 1 µSv/h above background (averaged over a 100m x 100m surface, or with a maximum spot dose in excess of 2.5 µSv/hr)[58],
- all unsafe buildings demolished and all contaminated materials either buried onsite or removed as dictated by risk,
- waste rock piles and the areas surrounding the open pit stabilized and adjusted as necessary to reduce environmental and safety hazard risks,
- any contaminants that pose an unacceptable environmental risk mitigated by containment onsite or removal,
- tailings areas stabilized and tailings removed or covered to prevent offsite migration and/or public exposure to unsafe contaminant levels.

These endpoints allow for traditional uses across the Gunnar site, although some specific areas such as tailings management sites, the open pit, and landfill sites may not be available for direct public uses such as camping or seasonal habitation. The endpoints would also preclude the location of any permanent structures such as cabins or cottages in or near the Gunnar site.

---

[58] This would allow for the area to be occupied continuously for up to 42 days before the maximum recommended annual dose level of 1,000 µSv would be reached.

# 9 A FINAL ACCOUNTING

Like all "one-industry" mining communities, Gunnar was completely dependent on the uranium industry, it boomed when the industry was "hot" and collapsed when the ore ran out. Such Cold-War-era boom and bust cycles were repeated in Canada, in nearby Uranium City [38,39,40,42] as well as in Elliot Lake, Ontario [25,200,221] and also in the United States, in such places as Uravan, Colorado; Moab, Utah; Jeffrey City, Wyoming; and Grants, New Mexico [7].

In the case of Gunnar Mines, like so many of the others, a natural resource was developed and exploited benefitting a local community, a region, and even a country with economic and national security benefits. On the other hand, these sites left behind a legacy of environmental disruption and damage, human and animal health risks, and fearsome clean-up costs.

## *Was it all worth it?*

There is no doubt that the Gunnar Mining operation provided huge local economic benefits for the Gunnar town-site, Uranium City, and the Athabasca basin communities, with lesser but still significant economic benefits to Saskatchewan and Canada. According to an evaluation of company reports, Gunnar Mines Ltd. produced 17.9 million pounds of $U_3O_8$ – comprising 6.89 million kg of uranium – and sold it for "a gross return of more than $147,000,000" (in 1964 Canadian dollars) [3]. Gunnar Mining Ltd.'s own final assessment in 1964 [168] was that total revenues derived from the sale of their uranium production between inception (August 1955) and closure (December 1963) came to $141.5 million. Taking Gunnar Ltd's. own valuation of about $142 million 1964 dollars, this amounts to about $1,100 million in 2016 Canadian dollars[59].

The Gunnar operation was profitable for its owners and produced taxes and royalties for the federal and provincial governments, respectively. Of the years for which the mine and mill were in production, the author was only able to locate financial statements for the years 1957 through 1963 (missing 1955 and 1956) so only an estimate can be provided here, although it should be reasonably indicative of the overall operation given that the available data cover 88% ($124.8 million of the $141.5 million total in revenue from uranium sales). With this caveat, the Gunnar Annual Reports' financial statements show that for the years 1957 through 1963 the revenue from uranium oxide sales was about $124.8 million, of which about $5.1 million was paid in royalties to the Saskatchewan government, about $6.0 million in taxes was paid to the Canadian government, with about $51.3 million in net profit (all in 1960s dollars as originally reported).

In addition, the principal driving purpose of the uranium exploration wave that led to the finding, development, and operations of Gunnar Mining was to find and develop uranium fuel for strategic military purposes in the cold-war era (see Section 1.4). This was surely a success as the uranium produced at Gunnar contributed substantially to the cold-war effort. When the Gunnar mine began production its output immediately doubled Canada's uranium production capacity (in 1955) and during its operating years the Gunnar mine was generally the second highest producer of the 16 Beaverlodge area mines in the Atomic Age and Cold War Eras (Section 1.4). How much uranium from the Gunnar mine ever found its way into nuclear weapons is either unknown or classified, but most of it would likely have found its way into such weapons, or the reserves for such weapons, or both. Gunnar's contributions to atomic bomb research and development were later applied to peaceful uses, such as nuclear medicine and nuclear power. Whether any or all of these uses of uranium amount to a net positive or negative benefit for society overall continues to be debated.

The other Gunnar legacy of a major uranium operation that was essentially just abandoned without remediation in 1964 (see Section 7.1), and then left idle for more than 40 years, has been a massive and expensive clean-up effort that continues as of the writing of this book.

The most recent projection made public by the Government of Saskatchewan is that the financial cost of the clean-up of the Gunnar mine, mill, and town site will be approximately $250 million [222,223]. This level of cost for a remediation of an operation of Gunnar's scale is not unique and has been experienced elsewhere. For example:

o In the United States, the Mi Vida uranium mine in Colorado

---

[59] $1 Canadian in 1964 would be worth $7.74 Canadian in 2016 according to "Inflation Calculator," http://inflationcalculator.ca/.

produced 12 million pounds of uranium ore during its operating life, "*enough to make at least eighteen atomic bombs*" [23]. The associated Utex mill remediation project has been estimated at U.S.$400 million [23].

o  In Australia, the Rum Jungle was a uranium deposit in the Northern Territory, Australia. Discovered in 1949, a mine and mill were constructed in 1952 that operated from 1953 to 1971. The initial 10-year project produced about 3.2 million pounds of uranium oxide [224]. Upon closure, the Australian government decided not to rehabilitate the mine site, as a result of which acid and metals leached into the nearby East Finniss River for many years. In addition, the abandoned open-pit mine was converted to a lake, which also became contaminated. After mining, the area suffered elevated gamma-ray radiation, alpha-ray emitting radioactive dust, and significant radon concentrations in air. Successive attempts to clean up the Rum Jungle site were made in 1977, 1983, 1990, and again in 2009, spending over A$25.7 million. In 2003, a government survey of the tailings piles at Rum Jungle found that capping which was supposed to help contain this radioactive waste for at least 100 years, had failed in less than 20 years. It has been estimated that the final remediation could cost an additional A$100-200 million [225].

Finally, another way to look at the cost-benefit aspect is a purely financial one: the value of the uranium produced by Gunnar is about $1,100 million (in 2016 Canadian dollars), and the latest public estimate of the total remediation cost is $250 million (also in 2016 Canadian dollars). Of course this does not include the costs of building and operating the Gunnar mine, mill, and town-site.

## *What have we learned?*

The Gunnar uranium mine story illustrates a marvel of engineering and industrial creativity given the geographical location and the period in history in which it was designed, built, and operated. The remediation of the Gunnar site provides an illustration of using modern science and engineering to provide multiple options for dealing with the hazards and the clean-ups of the aftermath of such industrial activities [144,179,216,226]. On the other hand, this story demonstrates that cleaning-up such legacy hazards from the past can be huge undertaking, and can be tremendously expensive. When not properly planned-for from the beginning, the remediation phase of such industrial development can end-up costing almost as much as the value of the original extracted resource. A key lesson is that mine and mill remediation and reclamation are best considered, planned-for, and budgeted-for at the beginning (before mining ever begins), as part of a comprehensive, full-cycle (sometimes referred-to as "cradle-to-grave") approach to uranium development.

# 10 GLOSSARY

**25** — A World War II-era code word for uranium-235. *See* Atomic Code Words.

**42-17 grade Z** — A World War II-era code word for uranium oxide. *See* Atomic Code Words.

**ACM** — Asbestos containing material. At Gunnar, the exteriors of most, if not all residences had 'transite type" asbestos-shingle siding made of a cemented ACM.

**Atomic Code Words** — During the World War II era atomic power research and uranium production were conducted in secrecy. In communications among partners Canada, the U.S., and U.K. code words were used to refer to materials such as uranium oxide (42-17 grade Z), uranium-235 (25), and heavy water (polymer) [25].

**AECB** — *See* Atomic Energy Control Board.

**AECL** — *See* Atomic Energy of Canada Ltd.

**Atomic Energy Control Board** — (AECB) An entity created in August 1946, under the Atomic Energy Control Act, to control and supervise *"the development, application and use of atomic energy"* [25]. The Board had wide regulatory authority that spanned research, mining, production, transportation, and use of prescribed "atomic substances" [25]. AECB was superseded by the Canadian Nuclear Safety Commission in 2000.

**Atomic Energy**
**of Canada Ltd.** (AECL) A Crown Corporation created in 1952 to assume responsibility for the nuclear research program formerly conducted by the National Research Council of Canada.

**Atomic Energy**
**Worker** The 1960 AECB regulations defined for the first time the concept of workers in jobs that could cause them to be exposed to nuclear radiation. AECB also regulated the maximum amounts of radiation to which such a worker could be allowed to become exposed. The AECB regulations also defined the maximum amounts of ionizing radiation to which a member of the general public could be allowed to become exposed, at $1/10^{th}$ of the amount for an atomic energy worker. In modern practice the term has become *"Nuclear Energy Worker (NEW)"* and is defined by the Canadian Nuclear Safety Commission.

**Atomic Pile** The first nuclear reactor cores contained a "pile" of layers of uranium pellets alternating with graphite bricks.

**Cage** A cage-like elevator car, suspended from a hoist on steel wire rope and used to transport miners and equipment up and down an underground mine shaft. Also called a Mine Cage. *See also* Skip.

**Canadian Nuclear**
**Safety**
**Commission** (CNSC) Canada's modern-day regulator, which regulates the use of nuclear energy and materials to protect health, safety, security and the environment. CNSC was established in 2000 to replace the former Atomic Energy Control Board.

**CNSC** *See* Canadian Nuclear Safety Commission.

**Decommissioning**
All technical and administrative actions leading to the release of a facility from regulatory control. This usually includes preliminary characterization, preparation and licensing of the strategy and activities, clean-up, decontamination, and dismantling activities, segregation and packaging of radioactive and non-radioactive wastes, and the final radiological monitoring for release. See [82].

**Drift** *See* Stope.

**Drillmaster Hammer Drill**  A pneumatic rock drill (also called drifter) that combined hammer drill and rotational motions to drill holes into rock while pushing the cuttings upwards to the surface for collection or disposal. Used at the Gunnar mine.

**Dygel**  *See* Forcite.

**EBR-I**  *See* Experimental Breeder Reactor I.

**Environmental Remediation**  Activities aimed at reducing radiation exposure from existing or potential contamination of land areas. This usually includes actions aimed at the contamination itself (by reducing and/or confining the source) and/or at the pathways for human and environmental exposure. See [82].

**EQC**  *See* NSEQC.

**Euclid Truck**  Euclid Co. dominated the off-road, heavy hauling truck market in North America in the 1950s and 60s, and their trucks were used at the Gunnar mine. In this era Euclid produced such trucks with haul capacities as large as 95 tonnes (105 tons).

**Experimental Breeder Reactor I**  (EBR-I) The United States' first electric-power generating nuclear reactor, which was built in Idaho and started-up in December, 1951.

**Forcite**  A "gelatin dynamite," comprising 30 to 80% nitroglycerin mixed with cellulose, sodium or potassium nitrate, and a hydrocarbon like tar (to make it waterproof). Dygel (a trademark of ICI Canada) seems to have been another gelatin dynamite formulation.

**Geiger-Müller Meter**  One of the first commercial hand-held radiation detector/counters. The Geiger-Müller Meter uses an ionization-chamber detector of the same name, enabling it to detect alpha particles, beta particles, and gamma rays. Modern versions are still available today.

**Glacial Flour**  A colloquial term sometimes applied to the Gunnar overburden material (comprising permafrost muskeg and glacial silt) when it was dried. Also termed Rock Flour.

**Mine Cage**  *See* Cage.

| | |
|---|---|
| **Mine Skip** | *See* Skip. |
| **Muck** | Rock, including both ore and waste rock, that has been blasted from a mine face. |
| **National Research Experimental Reactor** | (NRX) Canada's second nuclear reactor. It was built at Chalk River, Ontario and commenced operation in 1947. NRX was a 10 MW (later 42 MW) heavy-water-moderated research reactor. It was built and operated by the National Research Council until 1952 and thereafter by Atomic Energy of Canada Ltd. It was closed in 1993. *See also* National Research Universal Reactor and Zero-Energy Experimental Pile Reactor. |
| **National Research Universal Reactor** | (NRU Reactor) Canada's third nuclear reactor. It was built at Chalk River, Ontario and commenced operation in 1957. NRU is a 135 MW heavy-water-moderated research reactor. As one of Canada's national science facilities it is used to generate isotopes for medical diagnoses and/or treatments, to generate neutrons for the Canadian Neutron Beam Centre, and it is also used in the development of CANDU reactor fuels and materials. It is still in operation. *See also* National Research Experimental Reactor and Zero-Energy Experimental Pile Reactor. |
| **NEW** | Nuclear Energy Worker. *See* Atomic Energy Worker. |
| **NPD Reactor** | *See* Nuclear Power Demonstration Reactor. |
| **NRU Reactor** | *See* National Research Universal Reactor. |
| **NRX Reactor** | *See* National Research Experimental Reactor. |
| **NSEQC** | The Northern Saskatchewan Environmental Quality Committee, comprising representatives from the northern municipal and First Nation communities that are impacted by northern mining operations in Saskatchewan. |
| **Nuclear Energy Worker** | (NEW) *See* Atomic Energy Worker. |
| **Nuclear Power Demonstration Reactor** | (NPD Reactor) Canada's first electric-power generating nuclear reactor, which was built in Ontario and started-up in June, 1962. |

**Polymer**  A World War II-era code word for heavy water. *See* Atomic Code Words.

**Radon**  Radon is a chemical element that occurs naturally as a decay product of radium, which in turn is a decay product of uranium. As a result, radium and radon tend to be found wherever there is uranium. Radon poses human health concerns, not so much from radon itself, but from the alpha particles emitted from its decay products: polonium-210 and polonium-214, which can be adsorbed onto fine solid particles and/or small water droplets in the air, then inhaled, and then trapped in the lungs. In this case the alpha particles emitted can directly irradiate lung tissue, which can cause lung cancer [23,25].

**Riprap**  An erosion-resistant ground cover comprised of fairly large, loose, angular stones. Riprap is used to stabilize and protect shorelines and shoreline structures against erosion by waves or scour by ice.

**Raise**  A vertical, or nearly vertical, opening in an underground mine that leads from one level to another, and sometimes all the way to the surface.

**Rock Flour**  *See* Glacial Flour.

**Skip**  A bucket-like container, suspended from a hoist on steel wire rope and used to transport mined ore and waste rock up an underground mine shaft to the surface. Also called a Mine Skip. *See also* Cage.

**Saskatchewan Research Council**  (SRC) A research and technology organization incorporated as a Crown Corporation and owned by the Government of Saskatchewan. SRC conducts independent applied, research, development, demonstration, testing, and commercialization.

**SRC**  *See* Saskatchewan Research Council.

**Stope**  In underground mining a stope is the open passageway space that is left behind after the ore has been mined. A near-horizontal such passageway is termed a drift. Stoping refers to the removal of the ore from this space, and is practiced when the surrounding rock is stable enough not to collapse after the ore has been mined out.

| | |
|---|---|
| Tournatractor | A bulldozer with large rubber tires rather than tracks, enabling it to operate at greater speeds, and with greater mobility, than a tracked bulldozer. The units used at the Gunnar mine would have been built in the United States by R.G. LeTourneau Inc. (later LeTourneau-Westinghouse). |
| Transite | *See* ACM. |
| Tube Alloys | The code name for the secret atomic weapons development programs of the U.S., UK, and Canada, that were merged in 1943. |
| U-235 | The specific isotope of uranium (U) that is involved in sustainable nuclear fission. U-235 is naturally present only in very low concentrations, less than one percent, in the main uranium isotope, which is U-238. The numbers refer to the relative atomic mass of the element – atoms of U-238 have three more neutrons in them than do atoms of U-235. |
| Yellowcake | The final precipitated oxides of uranium that result from the milling of raw uranium ore using a leach process. Although often referred to as $U_3O_8$, for older processes this is a bulk-average approximation. Yellowcake from Cold War-era milling operations was usually a mixture of $UO_2$ and $UO_3$ with minor amounts of uranyl hydroxide and uranyl sulphate. |
| ZEEP Reactor | *See* Zero-Energy Experimental Pile Reactor. |
| Zero-Energy Experimental Pile Reactor | (ZEEP Reactor) Canada's first nuclear reactor and the world's first non-U.S. reactor. It was built at Chalk River, Ontario and commenced operation in 1945. ZEEP was a heavy-water-moderated reactor and was used to irradiate uranium to produce plutonium, and also to irradiate thorium to produce uranium-233. It was closed in 1970. *See also* National Research Experimental Reactor and National Research Universal Reactor. |

# 11 APPENDICES

| | | |
|---|---|---|
| 1 | Surface Plan of the Gunnar Site (1957) | 146 |
| 2 | Pictorial Drawing of the Site (1961) | 147 |
| 3 | Gunnar Mines as Represented in a 3D Computer Model | 148 |
| 4 | Gunnar Mines' Ore Flow Diagram (1961) | 149 |
| 5 | Process Flow Illustration of the Gunnar Crushing Plant (1959) | 150 |
| 6 | Process Flow Illustration of the Mill Circuit | 151 |
| 7 | Approximate Unit Conversions | 152 |

## Appendix 1. Surface Plan of the Gunnar Site (1957).

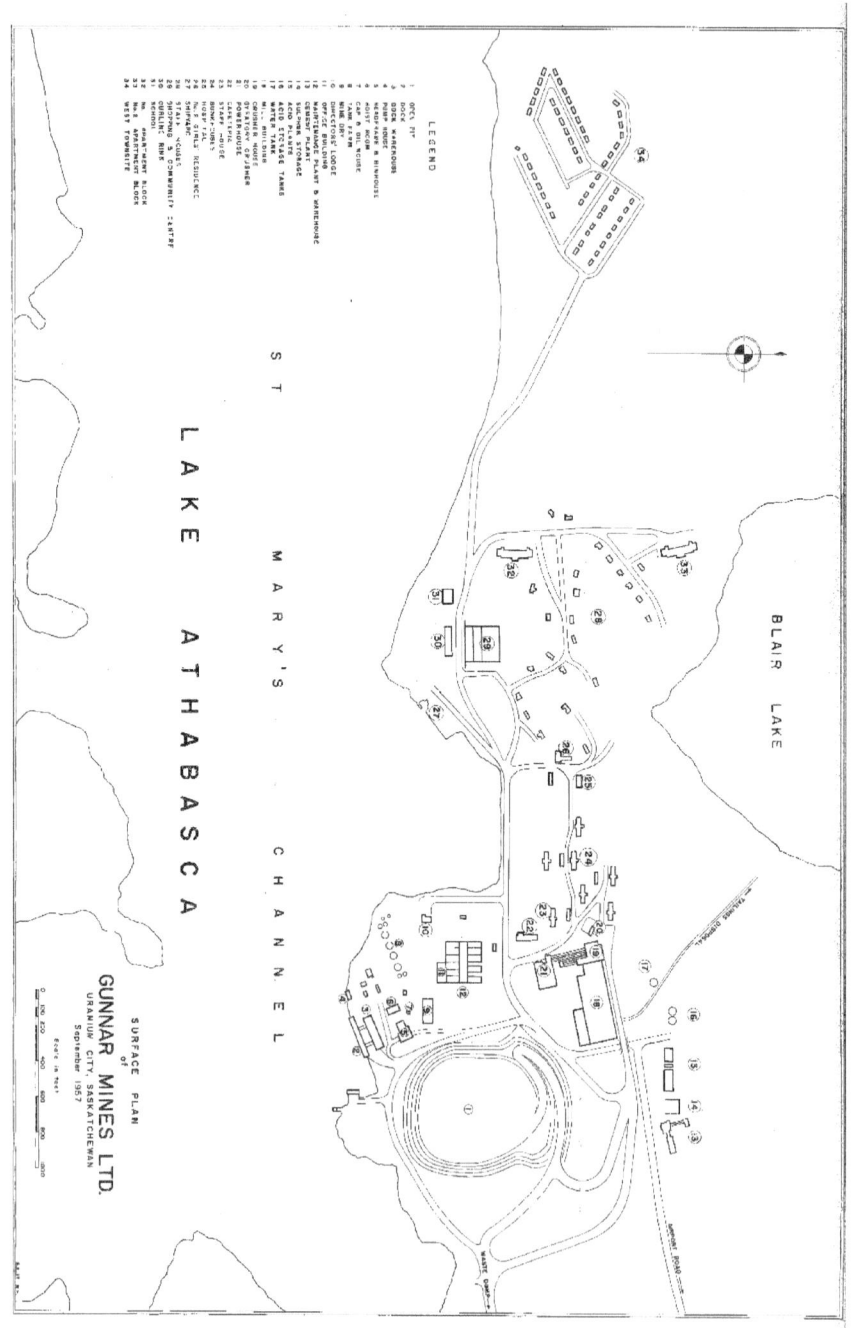

Appendix 2. Pictorial Drawing of the Site from 1961. (Gunnar Mines Ltd.).

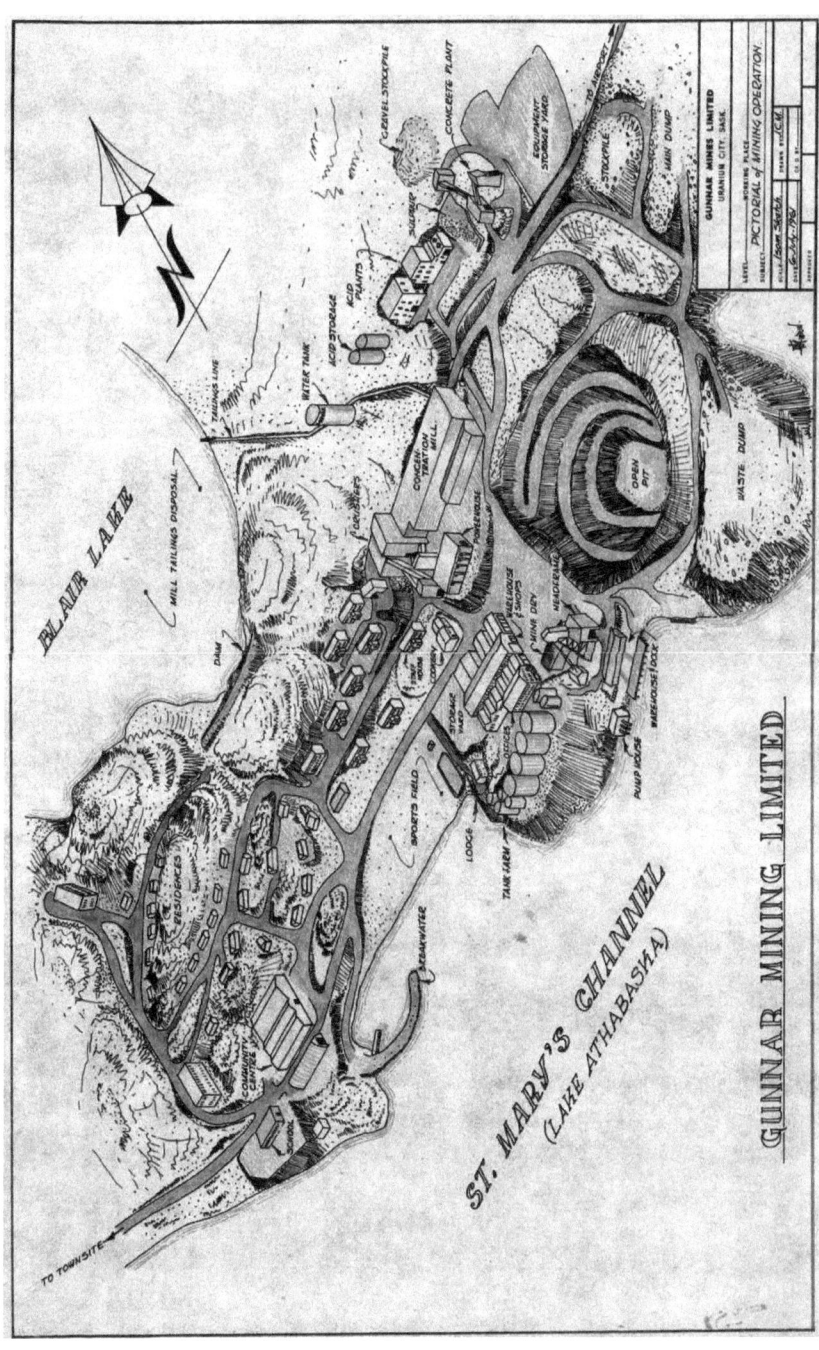

## Appendix 3. Gunnar Mines as Represented in a 3D Computer Model. Saskatchewan Research Council, 2016.

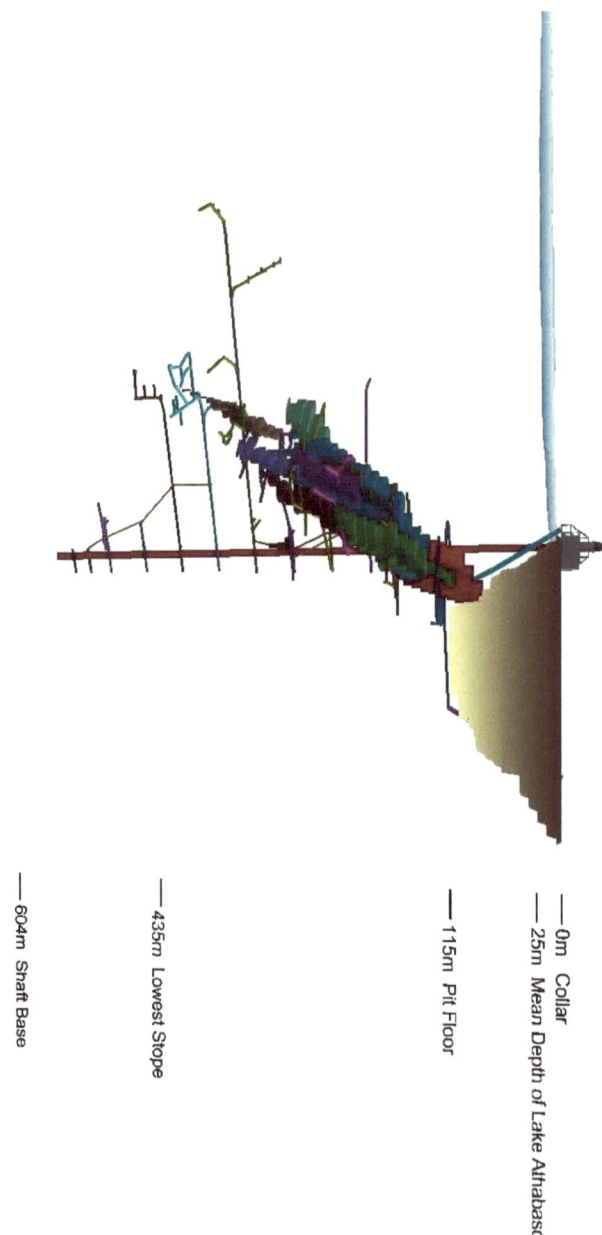

**Appendix 4. Gunnar Mines' Ore Flow Diagram. Gunnar Mines Ltd., July 5, 1961.**

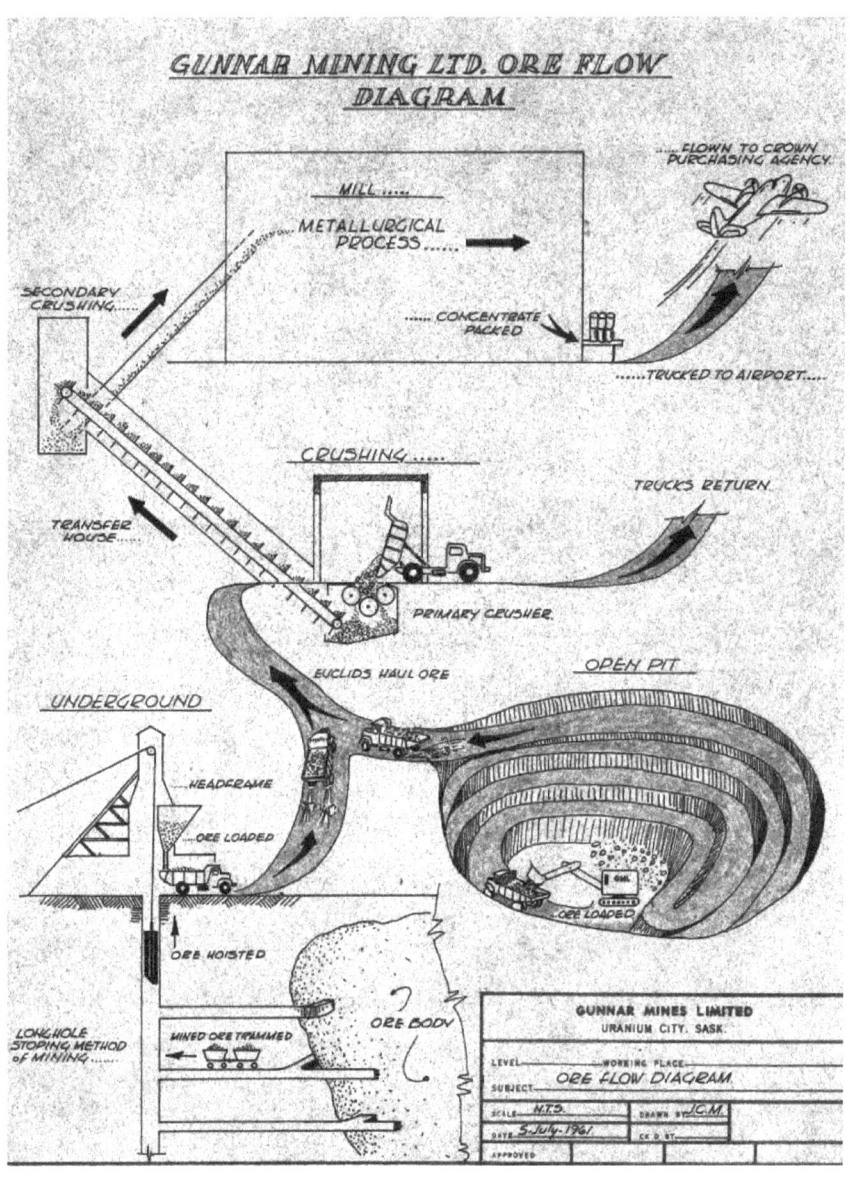

**Appendix 5. Process Flow Illustration of the Gunnar Crushing Plant (Gunnar Mines Ltd., May 6, 1959).**

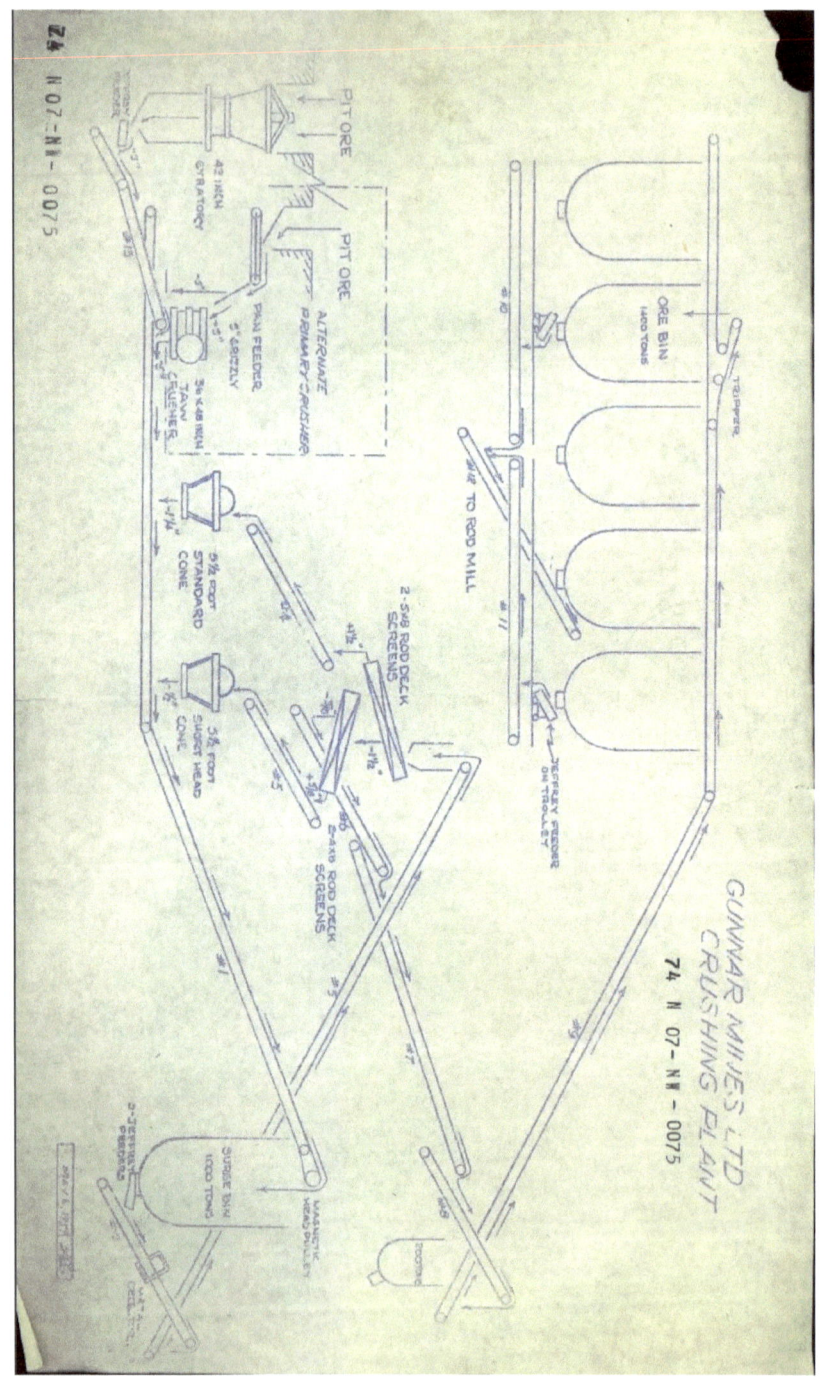

Appendix 6. Process Flow Illustration of the Mill Circuit (Gunnar Mines Ltd.).

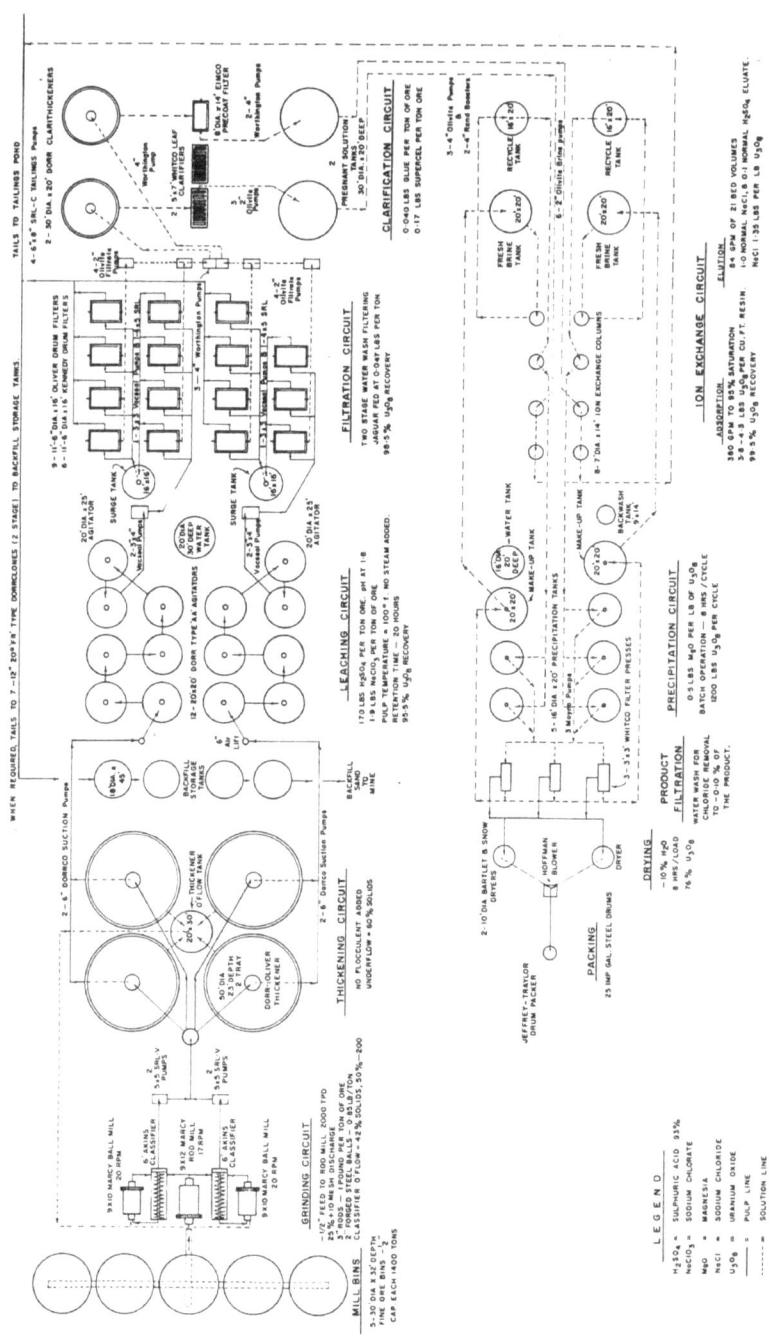

# Appendix 7. Approximate Unit Conversions.

These unit conversions are approximate only:

| Mass | Imperial pounds to kilograms | 1 lb = 0.454 kg |
|---|---|---|
| | Imperial tons to metric tonnes | 1 ton = 0.907 tonne |
| Distance | Imperial feet to metric metres | 1 ft = 0.3048 m |
| Volume | U.S. gallons to metric litres | 1 US gal = 3.785 l |
| | Imperial gallons to metric litres | 1 Imp gal = 4.546 l |

# 12 SUMMARY

## Gunnar Uranium Mine, Canada's Cold War Ghost Town

The Gunnar mine, mill, and town-site were built in a remote location in northern Saskatchewan, on the shore of Lake Athabasca. Like most mining communities the town boomed, first with construction workers and miners, and later with families. When the Gunnar mill construction was completed in the fall of 1955 it doubled Canada's uranium production capacity. By 1956 the Gunnar mine was the largest uranium producer in the world. The Gunnar town-site was built to serve the mine and mill and at one time had a population of about 850 people. By 1964 it was a ghost town. The Gunnar mine produced over 5 million tonnes of uranium ore, nearly 4.4 million tonnes of mine tailings, and an estimated 2,710,700 m3 of waste rock. Following closure in 1964, the Gunnar site was abandoned with little remediation and no reclamation being done. It has been referred-to as "the second greatest environmental disaster area in Canada." Forty years would pass before the governments of Saskatchewan and Canada reached an agreement to fund the remediation (clean-up) of the Gunnar site, and contracted the management of the project to the Saskatchewan Research Council (SRC). At the time of writing this book the clean-up was well underway, with several years of clean-up activity remaining, and a further expected 10-15 years of monitoring activity before the site is expected to be released into a long-term management and monitoring program.

**Print ISBN:** 978-0-9958081-2-6
**ePub ISBN:** 978-0-9958081-0-2

# 13 ABOUT THE AUTHOR

Dr. Laurier Schramm has over 35 years of R&D experience spanning each of the industry, not-for-profit, university, and government sectors. He is currently President and CEO of the Saskatchewan Research Council (SRC). His interests include technological innovation, management and leadership, colloid & interface science, and nanotechnology. He holds 17 patents, and has published 13 books and over 400 other publications and proprietary reports. He has served on many expert advisory panels and Boards, is co-founder of Innoventures Canada Inc. (I-CAN), and co-founder of Canada's Innovation School™. He has received national scientific and engineering awards for his work, and is a Fellow of the Chemical Institute of Canada and an honourary Member of the Engineering Institute of Canada.

**Conducting underwater reconnaissance at Gunnar in 2015.**

# 14 REFERENCES

1. UER.CA, "Uranium City," Urban Exploration Resource, UER.CA, 2014, http://www.uer.ca/locations/show.asp?locid=23608.
2. Moore, F., *Saskatchewan Ghost Towns*, First Impressions, Regina, SK, Sept. 1982.
3. Eldorado, *Uranium in Canada*, Eldorado Mining and Refining Ltd.: Ottawa, 1964.
4. Penrose, R.A.F., *Econ. Geol.*, **1915**, *10*, 161-171.
5. Hahn, O., "From the Natural Transmutations of Uranium to its Artificial Fission," In *Nobel Lectures, Chemistry 1942-1962*, Elsevier, Amsterdam, 1964, pp. 51-66 (Hahn's Nobel Prize Lecture given on 13 Dec. 1946).
6. Ringholz, R.C., *Uranium Frenzy, Boom and Bust on the Colorado Plateau*, Norton: NY, 1989.
7. Amundson, M., "*Yellowcake Towns: Uranium Mining Communities in the American West*," University Press of Colorado, Boulder, 2004.
8. Davidson, C.F., *The New Scientist*, **1957**, *(Feb. 21)* 9-11.
9. Taft, R.B., *Radium Lost and Found*, Furlong: Charleston, 1938.
10. Griffith, J.W., *The Uranium Industry - Its History, Technology and Prospects*, Mineral Report 12, Dept. of Energy, Mines and Resources: Ottawa, 1967.
11. Guidry, M., *The Guedry-Labine Family and the World's First Atomic Bomb*, accessed December 2013, http://freepages.genealogy.rootsweb.ancestry.com/~guedrylabinefamily/guedrylabineatomicbomb.
12. Bothwell, R., *Eldorado, Canada's National Uranium Company*, University of Toronto Press: Toronto, 1984.
13. Globe and Mail, "Gilbert LaBine: His Tools Were Pick, Paddle and .30-.30," *The Globe and Mail*, **1957**, *July 20*, p. 35.
14. Saskatchewan Department of Mineral Resources, *Inventory and Outlook of Saskatchewan's Mineral Resources*, Report No. 83, Dept. Mineral Resources:

Regina, SK, Nov., 1966, 52 pp.
15. CIM, *The Beaverlodge Uranium District*, Beaverlodge Branch, Canadian Institute of Mining & Metallurgy, Edmonton, Sept., 1957, 57 pp.
16. Saskatchewan Geological Survey, "Geology, and Mineral and Petroleum Resources of Saskatchewan 2003," Saskatchewan Industry and Resources: Regina, Misc. Report 2003-7, 2003.
17. Beck, L.S., "Uranium Deposits of the Athabasca Region," Report 126, Geological Survey, Saskatchewan Mineral Resources: Regina, 1969.
18. La Bine, D.G., *Gilbert A. LaBine 1890 – 1977*, accessed December 2013, http://www.labine.com/gilbert_a_labine, 2004.
19. Natural Resources Canada, *Atlas of Canada*, 6th Ed., Natural Resources Canada: Ottawa, 2009.
20. Hahn, O.; Strassmann, F. "Concerning the Existence of Alkaline Earth Metals Resulting from Neutron Irradiation of Uranium" *Naturwiss.*, **1939**, *27*, 11-15. Translation in *Am. J. Phys.*, **1964**, *January*, 9-15.
21. Meitner, L.; Frisch, O. R. "Disintegration of Uranium by Neutrons: a New Type of Nuclear Reaction," *Nature*, **1939**, *143 (3615)*, 239–240.
22. Peierls, R. "O. R. Frisch, 1904-1979," *Nature*, **1980**, *284 (13 March)*, 196–197.
23. Zoellner, T., *Uranium*, Penguin Books: London, 2009.
24. Rutherford, E., *Radio-Activity*, Cambridge University Press: Cambridge, 1904.
25. Sims, G.H.E., *A History of the Atomic Energy Control Board*, Canadian Government Printing Centre: Ottawa, 1980.
26. Dominion Bureau of Statistics, "Chronological Record of Canadian Mining Events from 1604 to 1943 and Historical Tables of the Mineral Production of Canada," Department of Trade and Commerce, Edmond Cloutier Printer: Ottawa, ON, 1945.
27. World Nuclear Association, *Brief History of Uranium Mining in Canada*, Appendix 1, World Nuclear Association: London, accessed January 2013, http://www.world-nuclear.org/info/Country-Profiles/Countries-A-F/Appendices/Uranium-in-Canada-Appendix-1--Brief-History-of-Uranium-Mining-in-Canada/.
28. "Early Instrumentation - 1920's," *National Radiation Instrument Catalog 1920 – 1960*, 2007, http://national-radiation-instrument-catalog.com/new_page_144.htm.
29. Taft, R.B., "Radium Hounds," *Scientific American*, **1939**, *160(1)*, 8-47.
30. LaBine, G.A., "Submission to Royal Commission on Canada's Economic Prospects," Government of Canada, Ottawa, 8 March 1956.
31. Hunter, W.D.G., "The Development of the Canadian Uranium Industry: An Experiment in Public Enterprise," *Can. J. Econ. Pol. Sci.*, **1962**, *28(3)*, 329-352.
32. Ross, M.; Hovdebo, D.G., "Uranium Mine Reclamation - A Myriad of Extremes Politics, Perceptions and Long-Lived Radionuclides," in Proc.19th Annual British Columbia Mine Reclamation Symposium, Dawson Creek, BC, pp. 188-196 (1995).
33. Athabasca Interim Advisory Panel, "Athabasca Land Use Plan: Stage

One," Ssakatchewan Environment: Regina, March 2006.
34. Hutton, E., "The Atom Bomb That Saves Lives," *Maclean's Magazine*, **1952**, *65(4)* February 15, p.7.
35. Fedoruk, S., "The Growth of Nuclear Medicine," *50 Years of Nuclear Fission in Review*, Canadian Nuclear Society: Ottawa, 1989, http://media.cns-snc.ca/history/fifty_years/fedoruk.html.
36. Idaho National Laboratory, "Experimental Breeder Reactor - I (EBR-I)," Brochure 07-GA50535_02, Idaho National Laboratory: Idaho, 2007.
37. Fawcett, R., *Nuclear pursuits: The scientific biography of Wilfred Bennett Lewis*, McGill-Queen's University Press: Montreal, 1994.
38. Belanger, D.; Hallett, F.; Dusseault, C., *The History of Uranium City*, Self-published. Available through several public libraries including the La Ronge Public Library and the Saskatoon Public Library, 1975, 19 pp.
39. Grade 10 Class Candu High School, *The History of Uranium City and District*, Lakeland Press, La Ronge, SK, 1982, 63 pp.
40. McBain, L., "Uranium City," Encyclopedia of Saskatchewan, University of Regina: Regina, 2006, http://esask.uregina.ca/entry/uranium_city.html.
41. Nichiporuk, A., "What Does the Future Hold for Uranium City?" *CIM Magazine*, **2007**, (November), 2 pp.
42. Keeling, A., "Born in an atomic test tube. Landscapes of cyclonic development at Uranium City: Saskatchewan," *The Canadian Geographer*, **2010**, *54(2)*, 228-252.
43. Northern Miner, Saskatchewan Uranium Shows Attract Monied Interest," *The Northern Miner*, 1952, Jan. 10, p. 1.
44. "The Uranium Rush - 1949," *National Radiation Instrument Catalog 1920 – 1960*, 2007, http://national-radiation-instrument-catalog.com/new_page_144.htm.
45. US Atomic Energy Commission, *Prospecting For Uranium*, US Government Printing Office: Washington, 1949.
46. Wright, R.J., *Prospecting with a Counter*, U.S. Atomic Energy Commission: Washington, 1954.
47. Northern Miner, "Uranium – Canada Maintains Place in Frantic World Production Race," *The Northern Miner*, 1952, Nov. 27, p.58.
48. Joubin, F.R.; James, D.H., "Canada's Uranium Future," *Precambrian*, **1956**, *29(5)*, 13-14.
49. Maclean's, "Uranium City Here We Come," *Reader's Digest Magazine*, **1954**, *64(384)*, April, 59-64.
50. Richardson, B.T., "The Hottest Square Mile in the World," Maclean's Magazine, **1951**, *64(20) Oct. 15, p. 14.*
51. Life, "Uranium Rush is On in Athabaska," *Life Magazine*, **1952**, *33(7)*, Aug. 18, pp. 15-19.
52. Stapleton, B., "Canada's Great Uranium Rush," *Collier's Magazine*, **1953**, *October 2*, pp. 32-41.
53. Advocate, "Atom Age Mining Rush Begins in N. Canada," *Advocate (Burnie, Tasmania)*, **1952**, *August 5*, p. 3.

54. Courier-Mail, "Began at dawn, Uranium rush in Canada," *The Courier-Mail (Brisbane, Queensland)*, **1952**, *August 5*, p. 1.
55. Sydney Morning Herald, "Canada's First Uranium Rush," *The Sydney Morning Herald (New South Wales)*, **1952**, *August 5*, p. 3.
56. Mercury, "Uranium rush in Canada," *The Mercury (Hobart, Tasmania)*, **1952**, *August 5*, p. 3.
57. TMC, "*The Birth of a Great Uranium Area*," Documentary Film, Technical Mine Consultants (TMC, Toronto) and Canadian Television Film Production, 1953.
58. Northern Miner, "Many Companies Active in Sask.," *The Northern Miner*, 1954, Sept. 16, p.2.
59. ITN, "*The Road to Uranium*," Documentary Film, Independent Television News (ITN), London, U.K., 16 October 1957.
60. MiningWatch Canada, Elliot Lake Uranium Mines, MiningWatch Canada: Ottawa, 2012, http://www.miningwatch.ca/elliot-lake-uranium-mines.
61. British Columbia Geological Survey, "MINFILE Mineral Inventory," MINFILE Record Summary, MINFILE No 082M 021, British Columbia Ministry of Energy and Mines: Victoria, BC, 2013, http://minfile.gov.bc.ca/Summary.aspx?minfilno=082M++021.
62. Kneen, J., "Uranium Mining in Canada – Past and Present," Presented to: *Indigenous World Uranium Summit*, Nov. 30-Dec. 1, 2006, Window Rock, Arizona. Accessed at: http://www.miningwatch.ca/sites/www.miningwatch.ca/files/Uranium_Canada_0/
63. Piper, L., *The Industrial Transformation of Subarctic Canada*, UBC Press: Vancouver, 2009.
64. Gunnar Mines Ltd., *The Gunnar Story*, Gunnar Mines Ltd., Toronto, Sept., 1957.
65. Delaney, G., "Uranium in Saskatchewan, Canada," Proc. South Australian Resources and Energy Investment Conference (SAREIC 2009), Unlocking South Australia's Mineral Wealth, 6 May 2009, http://www.pir.sa.gov.au/__data/assets/pdf_file/0006/104559/Gary_Delaney.pdf.
66. Tilman, A., "On the Yellowcake Trail," Parts 1-4, *Watershed Sentinel*, **2009**, *June-July*, 18-22; **2009**, *Sept.-Oct.*, 28-31; **2009**, *Nov.-Dec.*, 28-31; and **2009**, *Mar.-Apr.*, 28-31.
67. Piper, L., *Environment and History*, **2007**, *13*, 155-186.
68. Muldoon, J.A., "Policy Networks, Policy Change and Causal Factors, A Uranium Mining Case Study," Ph.D. Thesis, University of Regina, Regina, SK, March 31, 2015.
69. Schramm, L.L., *Research and Development on the Prairies. A History of the Saskatchewan Research Council*, Saskatchewan Research Council, Saskatoon, 2016.
70. Silversides, B.V.; Martin, S.F., "Cluff Lake, Portrait of a Canadian Mine," Cogema Resources Ltd.: Saskatoon, SK, 2002.
71. World Nuclear Association, "*Uranium in Canada*," World Nuclear

Association: London, U.K., January, 2016, http://www.world-nuclear.org/info/country-profiles/countries-a-f/canada--uranium.
72. Cameco, "Rabbit Lake," Cameco Corp.: Saskatoon, SK, accessed 27 July, 2016, https://www.cameco.com/businesses/uranium-operations/canada/rabbit-lake.
73. CNSC, "Comprehensive Study Report for Cluff Lake Decommissioning Project," Canadian Nuclear Safety Commission, Ottawa, December, 2003, 266 pp.
74. Cameco, "Key Lake," Cameco Corp.: Saskatoon, SK, accessed 31 December, 2013, http://www.cameco.com/mining/key_lake/.
75. Cameco, "Cigar Lake," Cameco Corp.: Saskatoon, SK, accessed 31 December, 2013, http://www.cameco.com/mining/cigar_lake/.
76. Bishop, C.S.; Goddard, G.J.H.; Mainville, A.G.; Paulsen, E., "Cigar Lake Project, Northern Saskatchewan, Canada," Technical Report, Cameco Corp., Feb. 24, 2012.
77. Cameco, "McArthur River," Cameco Corp.: Saskatoon, SK, accessed 31 December, 2013, http://www.cameco.com/mining/mcarthur_river/.
78. Bronkhorst, D.; Mainville, A.G.; Murdock, G.M.; Yesnik, L.D., "McArthur River Operation, Northern Saskatchewan, Canada," Technical Report, Cameco Corp.: Saskatoon, SK, Nov. 2, 2012.
79. Natural Resources Canada, "About Uranium," Natural Resources Canada, Ottawa, January, 2016, http://www.nrcan.gc.ca/energy/uranium-nuclear/7695.
80. World Nuclear Association, "Supply of Uranium," World Nuclear Association: London, U.K., August, 2012, http://www.world-nuclear.org/info/Nuclear-Fuel-Cycle/Uranium-Resources/Supply-of-Uranium/.
81. World Nuclear Association, "Supply of Uranium," World Nuclear Association: London, U.K., July, 2013, http://www.world-nuclear.org/info/Facts-and-Figures/Uranium-production-figures/.
82. Joint Research Centre, "Advancing Implementation of Nuclear Decommissioning and Environmental Remediation Programmes," Policy Support Document EUR 27902, European Commission, Brussels, 2016.
83. Zeemel, A., "How We Discovered and Staked the Gunnar Uranium Mine," *Precambrian*, **1956**, *29(5)*, 6-14.
84. Schiller, R., "Athabaska's Atom Boom," *Maclean's Magazine*, **1954**, *67(5)* Mar., 12-54.
85. Gunnar Gold Mines, "Nineteenth Annual Report. For the Year 1952," Gunnar Gold Mines Ltd., Toronto, 22 April, 1953.
86. Gunnar Mines, "20th Annual Report. For the Year 1953," Gunnar Mines Ltd., Toronto, 30 July, 1954.
87. Northern Miner, "Uranium Chances Spur Gunnar," *The Northern Miner*, 1952, Oct. 2, p.1,5.
88. Northern Miner, "Best Hole Yet for Gunnar," *The Northern Miner*, 1952, Oct. 16, p.1.
89. Northern Miner, "Intensify Drilling Program for Gunnar Uranium

Show," *The Northern Miner*, 1952, Nov. 6, p.1,5.
90. Northern Miner, "Uranium Project Suggests Big Tonnage – Gunnar Reports on Big Orebody," *The Northern Miner*, 1953, Feb. 26, p.1,16.
91. Northern Miner, "Gunnar Expands Ore Picture," *The Northern Miner*, 1953, Mar. 12, p.16.
92. Globe and Mail, "Gunnar Drilling Cuts Big Widths Of Uranium Ore," *The Globe and Mail*, **1952**, *Dec. 4*, p. 26.
93. Scott, J. "Ore Widths Large: Gunnar Drilling Shows Big Tonnage Potential," *The Globe and Mail*, **1953**, *Feb. 26*, p. 22.
94. Globe and Mail, "Gunnar Orebody Extended to West; Find North Limit," *The Globe and Mail*, **1953**, *Apr. 23*, p. 20.
95. Globe and Mail, "New tests show higher values on Gunnar group," *The Globe and Mail*, **1953**, *Jan. 12*, p. 21.
96. Globe and Mail, "Gunnar Gold Mines Canada's uranium bonanza," *The Globe and Mail*, **1954**, *Mar. 19*, p. 10, and *Mar. 22*, p. 30.
97. Scott, J., "Athabasca Uranium: Gunnar Kindles Blaze As Eldorado Ace Mine Nears Producing Stage," *The Globe and Mail*, **1953**, *Feb. 9*, p. 25.
98. Globe and Mail, "Five Producers in 1954: More Uranium Output Seen by Saskatchewan," *The Globe and Mail*, **1953**, *Dec. 24*, p. 15.
99. New York Times, "Uranium boom on in Saskatchewan," *New York Times*, **1953**, *May 28*, p. 41.
100. Scott, J., "Lake Cinch Is Newest Uranium Producer," *The Globe and Mail*, **1957**, *Sept. 21*, p. 41.
101. Northern Miner, "Gunnar Orebody Spreads West," *The Northern Miner*, 1953, Apr. 23, p.1,8.
102. Northern Miner, "Gunnar Orders 750-Ton Mill for its Uranium Property," *The Northern Miner*, 1953, June 11, p.17.
103. Northern Miner, "New Ore Shaping at Gunnar Gold," *The Northern Miner*, 1953, July 2, p. 1,8.
104. Life, "History's Greatest Metal Hunt," *Life Magazine*, **1955**, *38(21)*, May 23, p. 25-35.
105. Life, "The Uranium Rush is on in Athabaska," *Life Magazine*, **1954**, *33(7)*, Aug. 18, p. 15-19.
106. Argus, "Canada Looks to Active Uranium Year," *The Argus (Melbourne, Victoria)*, **1954**, *Jan. 9*, p. 17.
107. Courier-Mail, "Canada has a 'Rum Jungle'," *The Courier-Mail (Brisbane, Queensland)*, **1954**, *Aug. 3*, p. 2.
108. Northern Miner, "Big Freighting Job Major Time Saver in Building Gunnar," *The Northern Miner*, 1955, Oct. 13, p.28.
109. Northern Miner, "Record Tonnage Moves Down North to Uranium Mines," *The Northern Miner*, 1956, Oct. 4, p.17,20.
110. Northern Miner, "Gunnar Gold Drilling Enlarges Size of Main Uranium Ore Area," *The Northern Miner*, 1953, Aug. 27, p.1,6.
111. Northern Miner, "Gunnar Picture Still Growing," *The Northern Miner*, 1953, Oct. 29, p.1,4.
112. Northern Miner, "Gunnar Sets Objective for 1955 Production," *The Northern Miner*, 1954, Mar. 11, p.17,24.

113. Globe and Mail, "Elemental Sulphur," photo and caption, *The Globe and Mail*, **1954**, *Sept. 23*, p. 25.
114. Northern Miner, "Gunnar Mine Gets Set for Production – Aims for Start in Fall of Next Year," *The Northern Miner*, 1954, May 6, p.1,7.
115. Northern Miner, "Airlift gets Gunnar off to Early Start," *The Northern Miner*, 1954, May 20, p.17.
116. Northern Miner, "Make Progress at Gunnar Mine," *The Northern Miner*, 1954, June 17, p.1,8.
117. Northern Miner, "Gunnar's Uranium Plant Rapidly Taking Shape Mine," *The Northern Miner*, 1955, Jan. 27, p.17.
118. Gunnar Mines, "21st Annual Report. For the Year 1954," Gunnar Mines Ltd., Toronto, 13 April, 1955.
119. Northern Miner, "Gunnar Moves Waste Rock from Top of Orebody," *The Northern Miner*, 1955, May 26, p.3.
120. Northern Miner, "Gunnar Objective Will be Attained," *The Northern Miner*, 1955, June 2, p.2.
121. Northern Miner, "Gunnar Mine Looks Impressive as Production Draws Near," *The Northern Miner*, 1955, June 30, p.1,4.
122. Northern Miner, "Gunnar Mill Starts," *The Northern Miner*, 1955, Aug. 23, p.16.
123. Northern Miner, "Gunnar Mines in Production as Scheduled," *The Northern Miner*, 1955, Sept. 1, p.3.
124. Northern Miner, "First Precipitate Produced by Gunnar," *The Northern Miner*, 1955, Oct. 13, p.39.
125. Globe and Mail, "Population of Uranium City Now 930 and Still Growing," *The Globe and Mail*, **1955**, *Feb. 23*, p. 20.
126. Northern Miner, "Living is Pleasant in Gunnar Townsite," *The Northern Miner*, 1955, Oct. 13, p.37.
127. Northern Miner, "Gunnar's Shaft Mill Expansion Up to Schedule," *The Northern Miner*, 1956, July 12, p. 17.
128. Gunnar Mining Ltd., "The Gunnar Story," *Can. Mining J.*, **1963**, *7*, 53-119.
129. Gunnar Mines, "25th Annual Report. For the Year 1958," Gunnar Mines Ltd., Toronto, 23 March, 1959.
130. Gunnar Mining, "27th Annual Report. For the Year 1960," Gunnar Mining Ltd., Toronto, 3 March, 1961.
131. Gray, W., "No Beer Parlor for 7 Weeks: Uranium City Sheds Winter Gloom," *The Globe and Mail*, **1959**, *Apr. 1*, p. 13.
132. Gunnar Mining, "28th Annual Report. For the Year 1961," Gunnar Mining Ltd., Toronto, 6 March, 1962.
133. Hanion, M., "Freeze Brings Uranium Town Bleak Option," *Toronto Star*, **1962**, *Jan. 23*, p. A14.
134. Saskatchewan Research Council, KHS Environmental Management Group, and CanNorth Environmental Services, "Gunnar Site Characterization and Remedial Options Review," SRC Publication No. 11882-1C04, Saskatoon, January, 2005.
135. Beckett, T., "Gunnar Mines Part 1," *Uranium City History*,

http://uraniumcity-history.com/places/gunnar-part-1/, accessed September, 2014.
136. Northern Miner, "Gunnar's Mines Hitting Stride as Major Uranium Producer," *The Northern Miner*, 1956, Oct. 4, pp. 1,7.
137. Jackson, A.Y., "Gunnar Mines, 1957," graphite drawing, The National Gallery of Canada, Ottawa, 2016, http://www.gallery.ca/en/see/collections/artwork.php?mkey=41404.
138. Jackson, A.Y., "Gunnar Mines, Lake Athabaska, Saskatchewan, 1957," graphite drawing, The National Gallery of Canada, Ottawa, 2016, http://www.gallery.ca/en/see/collections/artwork.php?mkey=1312.
139. Gunnar Mines, *"Gunnar Progress,"* Documentary Film, Gunnar Mines Ltd., Toronto, *circa*. 1958.
140. Gunnar Mines, "26th Annual Report. For the Year 1959," Gunnar Mines Ltd., Toronto, 10 March, 1960.
141. Northern Miner, "Gunnar Shows Earning Power. July Output Over $1.6 Million," *The Northern Miner*, 1957, Aug. 15, pp. 1,7.
142. Chislett, W., "Storied Uranium City Losing Midas Touch; Exodus Expected Soon," *The Globe and Mail*, **1954**, *Sept. 21*, p. 7.
143. Nowicki, K. (Ed.), *1962 Memories*, school yearbook, Gunnar School, Gunnar, Sask., 1962.
144. Muldoon, J.; Schramm, L.L., "Gunnar uranium mine remediation project. Northern Saskatchewan," Proc. 33rd Arctic and Marine Oilspill Program (AMOP) Technical Seminar on Environmental Contamination and Response, Halifax, N.S., June 7-9, pp. 383-403, 2010.
145. Northern Miner, "Gunnar's $1,000,000 Power Plant," *The Northern Miner*, 1955, Oct. 13, p. 34.
146. Northern Miner, "Gunnar Profit Substantiated. To Start Underground Soon," *The Northern Miner*, 1957, May 2, pp. 17, 26.
147. Northern Miner, "Gunnar Profit Hits New Top" *The Northern Miner*, 1957, Oct. 24, pp. 1,16.
148. Quiring, D.M., *CCF Colonialism in Northern Saskatchewan: Battling Parish Priests, Bootleggers, and Fur Sharks*, UBC Press, Vancouver, 2004.
149. Dupas, R., "No. 1150. Douglas DC-3C (CF-GHX c/n 11780) Gunnar-Nesbitt Mines," Ron Dupas Collection, 1000aircraftphotos.com, 03/31/2011, http://1000aircraftphotos.com/Transports/1150.htm.
150. Webb, N., "North Atlantic Aviation Museum," Aviation Photos & History from Neville Webb, 2010, http://www.ruudleeuw.com/others-webb.htm.
151. Gandhi, S.S., "Age and Origin of Pitchblende from the Gunnar Deposit, Saskatchewan," in *Current Research Part B*, Paper 83-1B, Geological Survey of Canada: Ottawa, pp. 291-297, 1983.
152. Beck, L.S., "A Preliminary Report of Uranium Deposits of the Athabasca Region, Saskatchewan," Report 112, Geological Sciences Branch, Saskatchewan Mineral Resources: Regina, 1967.
153. Evoy, E.F., "Geology of the Gunnar Uranium Deposit Beaverlodge Area, Saskatchewan," Ph.D. Thesis, University of Wisconsin, 1961.
154. Ashton, K.E., "The Gunnar Mine: An Episyanite-Hosted, Granite-

Related Uranium Deposit in the Beaverlodge Uranium District" in *Summary of Investigations 2010*, Vol. 2, Saskatchewan Geological Survey: Regina, SK, Paper A4, 2010, 21 pp.
155. Evoy, E.F., "Open Pit Development at Gunnar," *Mining Engineering*, **1956**, *May*, 501-505.
156. Fraser, J.A.; Robertson, S.C., "Preliminary Description of the Geology and Mineralogy of the Gunnar Deposit, Saskatchewan," *Can. Min. Jour.*, **1954**, *75(7)*, 59-62.
157. Joliffe, A.W., "The Gunnar 'A' orebody," *CIM Trans.*, **1956**, 59, 181-185.
158. Evoy, E.F., "The Gunnar Uranium Deposit," in *Uranium Deposits of Canada*, Evans, E.L. (Ed.), Spec. Vol. 33, Can. Inst. Mining Metallurgy, Montreal, 1986, pp. 250-260.
159. Northern Miner, "Gunnar Sinks Shaft," *The Northern Miner*, 1955, Oct. 13, p. 32.
160. Northern Miner, "Gunnar Raises Sights for 1957 First Year Objective Exceeded," *The Northern Miner*, 1957, Jan. 3, p. 4.
161. Globe and Mail, "$20,000,000 Plant: Gunnar Is the Largest Open-Pit Uranium Mine," *The Globe and Mail*, **1955**, *Oct. 25*, p. 41.
162. New York Times, "Far Northern Uranium Mill to Open: Gunnar Mines Starts Operation 3 Years after Discovery," *New York Times*, **1955**, *Oct. 17*, p. 41.
163. Forrest, J.G., "Richest Atom Pit Begins Operation: Gunnar Mine in Production in Booming Beaverlodge," *New York Times*, **1955**, *Oct. 24*, p. 35.
164. Northern Miner, "Deep Work Expands Potential of Gunnar Underground Mine," *The Northern Miner*, 1958, Oct. 16, p. 8.
165. Gunnar Mining Ltd., "Pictorial, Gunnar Mining Limited Workings, Gunnar, Sask., Canada," Gunnar Mining Ltd., Toronto, June 30, 1961.
166. Northern Miner, "Gunnar Mines Hitting Stride as Major Uranium Producer," *The Northern Miner*, 1956, Oct. 4, p. 1.
167. Northern Miner, "Gunnar Now Solidly Entrenched as Winner of Handsome Profits," *The Northern Miner*, 1958, May 1, p. 1.
168. Gunnar Mining, "30th Annual Report. For the Year 1963," Gunnar Mining Ltd., Toronto, 3 April, 1964.
169. SRC, *12th Annual Report of the Saskatchewan Research Council 1958*, Saskatchewan Research Council, Regina, 1959.
170. Smithson, G.L.; Eager, R.L.; VanCleave, A.B., "Determination of Uranium in Flotation Concentrates and in Leach Liquors by X-Ray Fluorescence," *Can. J. Chem.*, **1959**, *37(1)*, 20-28.
171. Northern Miner, "Gunnar's Fine Mill to Reach Capacity Ahead of Schedule," *The Northern Miner*, 1955, Oct. 13, p. 34.
172. Northern Miner, "Marked Expansion Ahead for Uranium Mining," *The Northern Miner*, 1954, Nov. 25, p.49.
173. Northern Miner, "Sulphuric Acid Plant Really Important to Gunnar Process," *The Northern Miner*, 1955, Oct. 13, p. 33.
174. KHS, "Gunnar & Lorado. 2002-2003 Update," Report for

Saskatchewan Northern Affairs Dept., KHS Environmental Management Group Ltd., Dec., 2003.
175. Ruggles, R.G.; Robinson, D.J.; Zaidi, A., "A Study of Water Pollution in the Vicinity of Two Abandoned Uranium Mills in Northern Saskatchewan, 1978," Report EPS-MNR-5-81-2, Environment Canada, Ottawa, 1978.
176. Watters, R.; McKee, P.; Lush, D., "An Evaluation of Potential Environmental and Public Safety Impacts of Gunnar and Lorado Facilities in Northern Saskatchewan. Vol. 1. Summary of Existing Baseline Data," Report for Saskatchewan Environment by Beak Consultants Ltd., Sept., 1989.
177. BBT Consultants, "Gunnar Field Study," Prepared For Supply and Services Canada Under the National Uranium Tailings Program. NUTP No. - 155Q.2341-4-1674X (9 volumes). B.B.T. Geotechnical Consultants: IEC Beak Consultants Ltd; Sargent, Hauskins, Beckwith Concord Scientific Corp., March, 1986.
178. Kalin, M., "A Preliminary Assessment of the Environmental Conditions of Two Abandoned Uranium Mill Tailings Sites in Saskatchewan," Report EPS-5-WNR-81-1, Environment Canada, Ottawa, May, 1981.
179. Muldoon, J.; Schramm, L.L., "Gunnar Uranium Mine Environmental Remediation – Northern Saskatchewan," Paper ICEM2009-16102, Proc. 12$^{th}$ Internat. Conf. Environmental Remediation and Radioactive Waste Management - ICEM'09/DECOM'09, Liverpool, U.K., October 11-15, 2009.
180. Saskatchewan Research Council, "Former Gunnar Mining Limited Site Rehabilitation Project Proposal," SRC 12194-3E07, April, 2007.
181. Gunnar Mining, "29$^{th}$ Annual Report. For the Year 1962," Gunnar Mining Ltd., Toronto, 11 April, 1963.
182. SENES, "Screening Level Human Health and Ecological Risk Assessment for Gunnar Mine Site," Report for Saskatchewan Research Council, SENES Consultants Ltd., Richmond Hill, ON, March, 2006.
183. Brown, L.D., "Proposed Decommissioning of the Gunnar and Lorado Uranium Mine Sites," Report for Saskatchewan Environment, BB Health Physics Services, Regina, SK, 1993.
184. Gunnar Mining, "36th Annual Report. For the Year 1969," Gunnar Mining Ltd., Toronto, 5 June, 1970.
185. Northern Miner, "Gunnar, After Many Changes, Gone Forever," *The Northern Miner*, 1979, June 14, p. 25.
186. Tones, P.I., "Limnological and Fisheries Investigation of the Flooded Pit at the Gunnar Uranium Mine," Report C-805-10-E-82, Saskatchewan Research Council, Saskatoon, February, 1982.
187. Freshwater Fish Marketing Corp., "Annual Report. Year Ending April 30, 1981," Gunnar Mining Ltd., Winnipeg, 27 July, 1981.
188. Waite, D.T.; Joshi, S.R.; Sommerstad, H., "The Effect of Uranium Mine Tailings on Radionuclide Concentrations in Langley Bay, Saskatchewan, Canada," *Arch. Environ. Contam. Toxicol.*, **1988**, *17*, 373-380.
189. Stenson, R.; Howard, D., "Regulatory Oversight of the Legacy Gunner

Uranium Mine and Mill Site in Northern Saskatchewan, Canada – 13434," Proc., Waste Management Conference (WM2013), WM Symposia, Inc., Tempe, AZ, 2013, 13 pp.
190. CNSC, "Information and Recommendations of the Canadian Nuclear Safety Commission Staff in the Matter of Unlicensed Uranium Tailings Management Sites," CMD01-M77, Canadian Nuclear Safety Commission, Ottawa, 2001.
191. Saskatchewan, "Environment Audit of the Abandoned Gunnar and Lorado Mines Sites," Draft Report, Saskatchewan Environment, Regina, August, 1994.
192. Saskatchewan Research Council, "Gunnar Site Remediation Project: Environmental Impact Statement," SRC 12194-320-1L13, February, 2013.
193. Canada North Environmental Services (CanNorth), "Athabasca working group environmental monitoring program for the Athabasca communities year 2003," Prepared for the Athabasca Working Group, c/o Cameco Corporation, Saskatoon, Saskatchewan, 2004.
194. CanNorth, "Gunnar Site Characterization 2004 and 2005 Aquatic Assessments," Report for Saskatchewan Research Council, Project No. 1141, Canada North Environmental Services LP, Saskatoon, Saskatchewan, March, 2006.
195. Denison, "Demolition Strategy – Gunnar Uranium Mine, SK," Report for Saskatchewan Research Council, Denison Environmental Services, Elliot Lake, ON, March, 2006.
196. Fern, V., quoted in "For Our Children's Children", Band Brief, 2006. Referenced by "What's the Alternative to Nuclear Colonialism in the North?" Coalition for a Clean Green Saskatchewan, http://www.cleangreensask.ca/jim-harding/what%E2%80%99s-alternative-nuclear-colonialism-north, 16 October 2010.
197. Waggitt, P., "Uranium Mining Legacy Sites and Remediation - A Global Perspective," Presented at: IAEA Conference, Namibia, October, 2007, International Atomic Energy Agency, http://www.iaea.org/OurWork/ST/NE/NEFW/documents/RawMaterials/CD_TM_Swakopmund%20200710/13%20Waggit4.PDF.
198. IAEA, "Advancing Decommissioning and Environmental Remediation Programmes," IAEA Nuclear Energy Series Report No. NW-T-1.10, International Atomic Energy Agency, Vienna, 2016.
199. CNSC, "Uranium Mines and Mills Waste," Canadian Nuclear Safety Commission, Ottawa, 2014, http://www.nuclearsafety.gc.ca/eng/waste/uranium-mines-and-millswaste/index.cfm.
200. Larmour, A., "Elliot Lake Hailed as Reclamation Success Story," *Sudbury Mining Solutions J.*, **2010**, *Sept. 1*, http://www.sudburyminingsolutions.com/elliot-lake-hailed-as-reclamation-success-story.html.
201. AANDC, "Port Radium Mine (Remediation Complete)," Aboriginal Affairs and Northern Development Canada, Ottawa, 2012,

http://www.aadnc-aandc.gc.ca/eng/1332423218253/1332441057035.
202. SRC, "Project CLEANS (Cleanup of Abandoned Northern Sites)," Saskatchewan Research Council, Saskatoon, 2014, http://www.src.sk.ca/about/featured-projects/pages/project-cleans.aspx.
203. Castrilli, J.F., "Wanted: A Legal Regime to Clean Up Orphaned /Abandoned Mines in Canada," *J. Sust. Devel. Law Policy* **2010**, *6(2)*,109-141.
204. SENES, "Screening Level Human Health and Ecological Risk Assessment for Gunnar Mine Site," SENES Consultants Ltd., Richmond Hill, ON. Prepared for Saskatchewan Research Council, 2006.
205. Keeling, A., "Atomic Outpost," *Canada's History*, **2011**, *91(3)*, 28-35.
206. Peach, I.; Hovdebo, D., "Righting Past Wrongs: The Case for a Federal Role in Decommissioning and Reclaiming Abandoned Uranium Mines in Northern Saskatchewan," Public Policy Paper 21, Saskatchewan Institute of Public Policy, University of Regina, December, 2003.
207. Saskatchewan, "Canada's New Government and Province of Saskatchewan Launch First Phase of Cleanup of Legacy Uranium Mines," News Release, Government of Saskatchewan, Regina, April 2, 2007.
208. Editorial Board, "U.C. Meeting on Abandoned Mines," Supplement, *Opportunity North*, **2008**, *Spring*, 2-3.
209. Editorial Board, "Cleanup Process Begins at Gunnar," *Opportunity North*, **2009**, *Summer*, 22.
210. Editorial Board, "Project CLEANS Team Gears up for a New Work Season," *Opportunity North*, **2013**, *Spring*, 29 (see also p. 20).
211. Petelina, E., Sanscartier, D.; MacWilliam, S.; Ridsdale, R., "Environmental, Social, and Economic Benefits of Biochar Application for Land Reclamation Purposes," *Proc. 38$^{th}$ Ann. B.C. Mine Reclamation Symposium*, 2014, 13 pp.
212. Saskatchewan, "Northern Saskatchewan Environmental Quality Committee," Government of Saskatchewan, Regina, 2016, https://www.saskatchewan.ca/residents/first-nations-citizens/saskatchewan-first-nations-metis-and-northern-initiatives/northern-saskatchewan-environmental-quality-committee.
213. Provost, K., "Abandoned Mine Being Cleaned-Up in Saskatchewan," CJLR-FM News, La Ronge, SK, broadcast 25 October 2010.
214. SRC, "Gunnar Mine Rehabilitation Project, Structural Safety Assessment of Buildings and Other Structures," Saskatchewan Research Council, Saskatoon, September, 2010.
215. SRC, "Gunnar Mine Rehabilitation Project, Demolition Plan," Saskatchewan Research Council, Saskatoon, October, 2010.
216. Muldoon, J.; Yankovich, T., and Schramm, L.L., "Gunnar Uranium Mine Environmental Remediation - Northern Saskatchewan," Paper, ICEM2013-96223, *Proc. 15th Internat. Conf. on Environmental Remediation and Radioactive Waste Management, ICEM 2013*, Brussels, Belgium, Sept. 8-

12, 2013, 10 pp.
217. CNSC, "CNSC Accepts Environmental Assessment Report and Issues Licence for the Gunnar Remediation Project," News Release, Government of Canada, Ottawa, 15 January 2015, http://news.gc.ca/web/article-en.do?nid=920849.
218. SRC, "Gunnar Site Remediation Project – Tailings Remediation Plan," Saskatchewan Research Council, Saskatoon, August, 2015.
219. CNSC, "CNSC Removes the Gunnar Remediation Project Phase 2 Hold Point as it Pertains to the Remediation of the Tailings Deposits," News Release, Government of Canada, Ottawa, 27 November 2015, http://news.gc.ca/web/article-en.do?nid=1022169.
220. Redmann, R.E.; Frankling, F.T., "Revegetation of Abandoned Uranium Mill Tailings Near Uranium City, Saskatchewan. Plant Species Selection," Report for Saskatchewan Environment, Regina, March, 1982.
221. Mawhiney, A-M.; Pitblado, J. (Eds.), Boom Town Blues: Elliot Lake, Collapse and Revival in a Single Industry Community, Dundurn Press, Toronto, 1999.
222. MacPherson, A. "Gunnar Cleanup to Exceed $250M, 10 Times Estimate," *Saskatoon StarPhoenix,* October 17, 2015, Last Updated: February 18, 2016.
223. MacPherson, A. "Overbudget Gunnar Cleanup Federal Responsibility, Sask. Politicians Say," *Saskatoon StarPhoenix*, February 24, 2016.
224. AAEC, "Rum Jungle Project," Booklet, Australian Atomic Energy Commission, Lucas Heights, Australia, 1963.
225. World Nuclear Association, "Former Australian Uranium Mines," World Nuclear Association, London, U.K., 2014, http://www.world-nuclear.org/info/Country-Profiles/Countries-A-F/Appendices/Australia-s-former-uranium-mines/
226. Schramm, L.L. (2012) Cleaning-Up Abandoned Uranium Mines in Saskatchewan's North, (W. B. Lewis lecture) *Bulletin of the Canadian Nuclear Society*, **33**(2), 17-23.

www.ingramcontent.com/pod-product-compliance
Lightning Source LLC
Chambersburg PA
CBHW041619220426
43661CB00046B/1503